Lucent Technolo
Bell Labs Innov.

MW01487089

INDUSTRIAL ECOLOGY
AND
THE AUTOMOBILE

Thomas E. Graedel

Bell Laboratories, Lucent Technologies

*Now at School of Forestry and Environmental
Studies, Yale University*

Braden R. Allenby

AT&T

PRENTICE HALL
Upper Saddle River, New Jersey 07458

Library of Congress Cataloging-in-Publication Data

Graedel, Thomas E.
    Industrial ecology and the automobile / Thomas E. Graedel,
Braden R. Allenby
       p.   cm.
    Includes bibliographical references and index.
    ISBN 0-13-607409-X
    1. Automobiles—Design and construction.   2. Industrial ecology.
    3. Green technology.   I. Allenby, Braden R.   II. Title.
    TL240.G73   1997
    629.2'31—DD21   97–7761
                         CIP

Acquisitions Editor: **WILLIAM STENQUIST**
Editor-in-chief: **MARCIA HORTON**
Project editor: **IRWIN ZUCKER**
Managing editor: **BAYANI MENDOZA DE LEON**
Director of production and manufacturing: **DAVID W. RICCARDI**
Copy editor: **MARJORIE SHUSTAK**
Cover director: **JAYNE CONTE**
Manufacturing buyer: **JULIA MEEHAN**
Editorial assistant: **MARGARET WEIST**

 © 1998 by Bell Laboratories, Lucent Technologies
Published by Prentice-Hall, Inc.
Simon & Schuster/A Viacom Company
Upper Saddle River, New Jersey 07458

The author and publisher of this book have used their best efforts in preparing this book. These efforts include the
development, research, and testing of the theories and programs to determine their effectiveness. The author and
publisher shall not be liable in any event for incidental of consequential damages in connection with, or arising out
of, the furnishing, performance, or use of these programs.

Cover illustration: Hybrid vehicle from *Science and Technology Review* (CRL-5200-95-7), page 26. This illustration was prepared to document work done under the auspices of the University of California, Lawrence Livermore
National Laboratory, and the Department of Energy.

Printed in the United States of America

10 9 8 7 6 5 4 3 2 1

ISBN 0-13-607409-X

Prentice-Hall International (UK) Limited, London
Prentice-Hall of Australia Pty. Limited, Sydney
Prentice-Hall Canada Inc., Toronto
Prentice-Hall Hispanoamericana, S.A., Mexico
Prentice-Hall of India Private Limited, New Delhi
Prentice-Hall of Japan, Inc., Tokyo
Simon & Schuster Asia Pte. Ltd., Singapore
Editora Prentice-Hall do Brasil, Ltda., Rio de Janeiro

To
Susannah and Carolyn

Contents

Preface

The interactions of the societal-industrial system with the environment form one of the most critical issues in today's world. The inadequacy of current environmental regulatory structures and traditional ways of analyzing environmental issues, and the continuing need to mitigate the environmental perturbations arising from this complex relationship, have led to the development of a new conceptual framework termed *industrial ecology*.

Industrial ecology is the means by which humanity can deliberately and rationally approach and maintain a desirable carrying capacity, given continued economic, cultural, and technological evolution. The concept requires that an industrial system be viewed not in isolation from its surrounding systems, but in concert with them. It is a systems view in which one seeks to optimize the total materials cycle from virgin material, to finished material, to component, to product, to obsolete product, and to ultimate disposal. Factors to be optimized include resources, energy, and capital.

One of the industries of most interest to industrial ecology is that of the automobile. In large measure this is true because automobiles are so much a part of the lives of people in the world's more developed countries: they are essential transportation, they are status symbols, they are statements of maturity, they are icons of personal freedom, they are engines of national economies. They are also demonstrable stresses on the environment, as easily witnessed by their tailpipe emissions, the smog that forms from these emissions, and the difficulty of dealing with dirty oil or worn-out tires. These visual impressions contrast with the automotive's industry's often exemplary progress at dealing with environmental issues. Manufacturing facilities are much cleaner, emissions smaller, and recycling more completely accomplished than is the case with any other product of similar size and complexity.

All of these factors make the automobile and its environmental attributes an ideal case study in industrial ecology. In such a case study, one can explore experiences in designing the automobile with the environment in mind, now a focus of every automotive design team. What choices are desirable? Which decisions work? Which are difficult to implement? Where is more research needed? These are some of the topics that we treat in this book.

Although this book's discussions are centered on the automobile, they can be viewed as appropriate to all of technology. Most students of this book will not become automotive designers, but may well design other products: washing machines, building cooling systems, machine tools, infrastructure projects, even software systems. The automobile thus serves here as the introduction to more general Design for Environment, a key component of the repertoire of the 21st century engineer.

We hope this book will have several uses. First, we anticipate that it may serve as a module in the "introduction to engineering" courses now common in universities. Second, we hope it is useful as a case study in design courses in automotive, mechanical, and civil engineering classes. Third, we hope that students of societal systems may find some value in its discussion of the way in which industrial activities are intertwined with societal trends and with infrastructures. Fourth, we anticipate that the topic may be of interest to students of business and management, now striving to blend the challenges of business and the world within which it functions. To serve these diverse audiences, both technology-oriented and policy-oriented exercises are included in several of the chapters. Finally, practicing engineers may find the book of value as they continue to improve the environmental responsibility of their product, process, and infrastructure designs.

The metric system is used throughout the book; where necessary we have converted English units to metric. The latter is clearly the language of modern technology and its use here merely recognizes modern usage practice. An appendix presents conversion units for the convenience of the reader.

Industrial ecology is a broad subject whose many attributes can only be touched upon here. Thus, we mention in the book only in passing a number of topics that profit from more detailed discussion: technological change, risk assessment, economic and legal constraints, corporate structure, governmental interests, and the like. For those interested, we refer you to our full-length textbook *Industrial Ecology* (T. E. Graedel and B. R. Allenby, Prentice Hall, Inc., Englewood Cliffs, NJ, 1995).

We thank the following for providing information, diagrams, and/or reviews of drafts of *Industrial Ecology and the Automobile*: Robert Baboian (Texas Instruments, Inc.), Frank Field (Massachusetts Institute of Technology), Wayne France (General Motors Corporation), Robert Frosch (Harvard University), Chris Hendrickson (Carnegie Mellon University), Inge Horkeby (Volvo Car Corporation), Arpad Horvath (Carnegie Mellon University), Gregory Keoleian (University of Michigan), Sue McNeil (Carnegie Mellon University), Agneta Wendel (Volvo Car Corporation), and Ronald Williams (General Motors Corporation). Their assistance has resulted in a much improved product.

T. E. Graedel
B. R. Allenby

PART I: AUTOMOBILES AND SOCIETAL STRUCTURES

Humanity
on the
Move

A car is more than a car: It reflects the priorities of an era, the country, the people that shape it. It also contributes to defining those priorities. It is both chicken and egg, leader and follower. More than any other common product, the car is three-dimensional shorthand for who and what you are.

John Butman, *Car Wars*, Grafton Books, London, 1991

In 1989, the post-war totalitarian empire of the Soviet Union collapsed. Although its single-minded focus on economic and technological growth had created substantial environmental degradation, it had at least left as a legacy in many urban areas world-class mass transit systems. Prague's subway system, for example, is capable of moving 50,000 people an hour and has been highly praised by Western experts. So what happened in the six years after the fall of the empire? According to the March 7, 1995 Wall Street Journal, the Czechs now own 2 million private automobiles, up 500,000 in six years, while in Prague itself private automobile ownership is up by 45 percent. In the newly liberated Eastern European countries, the Journal notes, the automobile is still "an icon of liberty."

In the U.S., meanwhile, motor vehicles are estimated to contribute at least 50 percent of all carbon monoxide emissions, 29 percent of all nitrogen oxides emissions, and 27 percent of all volatile organic carbon (VOC) emissions (see Fig. 1.1). This has occurred despite the fact that, since controls were introduced in 1968, total VOC and CO emissions per vehicle have been reduced by some 96 percent, and, since imposition of NO_x controls in 1972, emissions of those species have been reduced by over 75 percent. However, although the U. S. population increased by 23 percent between 1970 and 1990, the population of automobiles increased from 108 million to 189 million, or by 75 percent. Moreover, during the same period the distance that the average vehicle was driven grew by 11 percent, and the mix of vehicles changed from sedans to more inefficient vans and trucks. The four

1

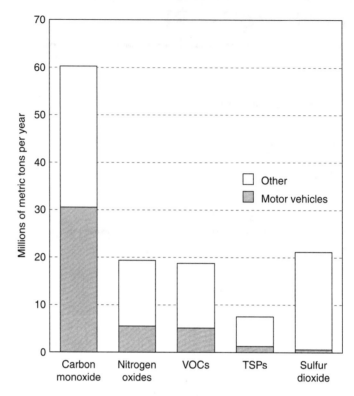

Figure 1.1   Motor vehicle contributions to U.S. air pollution emissions, 1990. VOC is volatile organic carbon compounds, TSP is total suspended particulates. (J.J. MacKenzie, *The Keys to the Car*, Washington, D.C.: World Resources Institute, 1994.)

wheel drive off-road recreational vehicles, which, as the March 18, 1995 issue of *The Economist* notes, are in many cases "really just fashion accessories," are particularly heavy and inefficient.

In part, these demand patterns are the result of the fact that the real price of gasoline in the United States is currently lower than at any time since the early 1920's: The cost of gasoline per kilometer in 1985 cents has dropped from 6.9 in 1980 to 3.1 in 1990, to 2.4 in 1994. As a consequence of these factors, transportation is the only major petroleum-consuming U.S. industrial sector where consumption has increased, not decreased, in the past decades (Fig. 1.2).

Another significant impact of the automobile arises from the use of petroleum as the principal motor vehicle fuel. As a concentrated energy source and the feedstock for the world's chemical and plastics industries, petroleum is a crucial and limited resource whose use should be carefully considered. However, something like half of all petroleum that is refined from crude oil extracted from its subterranean reservoirs is immediately consumed by the global motor vehicle fleet, and that proportion is increasing. Further, lead compounds are often added to automotive fuel to increase performance and to lubricate the

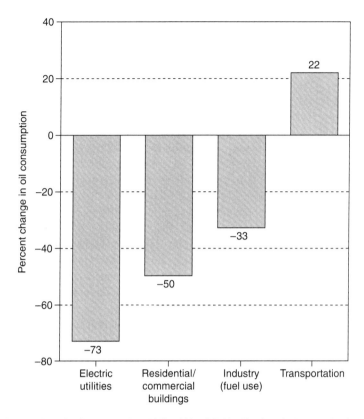

Figure 1.2   Changes in U.S. oil consumption, 1973–1992. (J.J. MacKenzie, *The Keys to the Car*, Washington, D.C.: World Resources Institute, 1994.)

engine during operation. Although the use of such fuel has been banned in a number of countries, it is still quite common around the world, and contributes to substantial dispersive use of this toxic heavy metal, frequently beyond national borders.

Even though these resource depletions and air emissions are important costs to the operation of private automobiles, they may not be the most critical. The automobile is not simply an object, it is also a technology system that has spawned enormous infrastructure and ancillary industrial structures. The growth of the modern petroleum industry is one not inconsequential result. A vast network of pipelines, oil tankers, refineries, product distribution networks, and, at the customer end, gasoline stations with literally millions of underground tanks around the world have been created to service the automobile. Millions of tons of aggregate, asphalt and concrete have been laid over hundreds of thousands of hectares of farms, forests, fields, and wetlands over the years to provide the road networks, parking lots, malls, and other elements of a transportation network largely defined by the private automobile. The concomitant siltation of streams and rivers during construction, and subsequent ongoing runoff—often containing heavy metals and other contaminants—are predominant non-point sources of water degradation. Moreover, because of the complexity of this automotive infrastructure, control of resulting environmental impacts has proven to be very expensive and politically sensitive.

More subtly, in many countries the dominance of the private automobile has led to a spatial distribution of population and employment in many localities that, in its current form, renders mass transit uneconomical because of the diffusion of demand away from defined corridors and specific major destinations. This pattern embeds a continuing need for personal transportation—the automobile—in many cultures for the forseeable future. Objectively, this is a substantial constraint to any proposal to eliminate the automobile, or even greatly reduce its use, in an effort to mitigate its environmental impacts in the short term. It puts a premium, however, on technological efforts to improve the environmental performance of the automobile itself as rapidly as possible, a goal to which this text is largely devoted.

The automotive technology system in virtually every society where it is common also illustrates an important principle first discussed nearly three decades ago by Garrett Hardin of the University of California, Santa Barbara. Hardin described what he termed "The Tragedy of the Commons"; its principal argument was that a society that permitted perfect freedom of action in activities that adversely influenced common properties was eventually doomed to failure. Hardin cited as an example a community pasture area, used by any local herdsman who chooses to do so. Each herdsman, seeking to maximize his financial well-being, concludes independently that he should add animals to his herd. In doing so, he derives additional income from his larger herd but is only weakly influenced by the effects of overgrazing, at least in the short term. At some point, however, depending on the size and lushness of the common pasture and the increasing population of animals, the overgrazing destroys the pasture and disaster strikes all.

Few people have herds of animals these days, but nearly everyone participates with great frequency in a modern version of the tragedy of the commons: the traffic jam. In this variation on the theme, the convenience, privacy, and safety of travel by private automobile encourages each individual to drive to work, school, or stores. At low levels of traffic density, this is a perfectly logical approach to the demands of modern life. At some critical density, however, the road network commons is only marginally able to deal with the traffic load, and the smallest disruption (a stalled vehicle, a delivery truck, a minor accident) dooms drivers to minutes or hours of idleness, the exact opposite of what they had in mind. Examples of frequent clogging of road network commons systems are now legendary: Los Angeles, Tokyo, Naples, Bangkok, and on and on.

One cannot help asking such questions as, "Can't we avoid this extensive and aggravating waste of time somehow?" "Will things only get worse?" "Is this the purpose for which our society has embraced automobiles so universally and uncritically?" And, perhaps the most crucial, "We can see how we got to the stage we are now in, but where do we want to get to, and can we be motivated to do it once we do know?" Note that these questions all contain two different, yet critical dimensions: a technological dimension, wherein technological solutions can act to achieve a desired goal, and a social dimension, which is frequently more fundamental, and more difficult to analyze and influence.

And yet there are other aspects to the automobile that are environmentally desirable. No other common economic article is repaired and refurbished as well as the automobile and its components: strong secondary markets exist for automotive components of all sorts, as well as for the automobile itself. Moreover, once its useful life is ended, the constituent materials in automobiles are widely recycled in the U.S. and most developed countries. About 95 percent of all automobiles enter the recycling system at the end of their useful life, and some 75 percent by weight of all materials used in each of these automobiles is recycled, with considerable research already underway on how that figure may be improved. The automobile can thus almost be thought of as a case study of successful recycling of a complex product, and as an example of product life extension. The latter is particularly interesting in that older automobiles as a class generate much greater emissions than newer models, so that extending their life without substantial modification may be, in fact, environmentally harmful. Of equal interest is the recent finding that the greatest class of polluting automobiles consists of those that are either poorly maintained or tampered with, thus linking the environmental performance of the automotive fleet as a whole directly to human behavior.

The study of automobiles from an industrial ecology perspective brings still another topic into the discussion: the infrastructure that support them—roads, bridges, repair shops, petroleum distribution systems, parking lots. As with many other products of our industrialized society, the automobile is best perceived as a focus of a technological system. Inevitably, the decisions of automotive design engineers influence not only the automobile but its infrastructure as well.

The automobile is, in short, a supremely instructive technology system for industrial ecologists. Few articles cause so much environmental impact, yet are so deeply embedded in modern culture and economies—so much so that imagining life without the automobile is, for most people, impossible. The difficulties and challenges of evolving psychological, cultural and social attitudes towards sustainability are seldom so obvious as they are with the car. Germany, for example, is one of the most environmentally-conscious countries in the world, with a large and politically powerful Green Party, yet it still has no speed limits on the autobahn. Similarly, technological challenges are well illustrated by the automobile, especially as regards design tradeoffs arising in different lifecycle stages of the automobile. Moreover, few technology systems are so pervasive, have so many different and significant environmental impacts, and thus offer such substantial potential for environmental improvement.

In discussing this remarkable industrial sector, we will focus on three important industrial ecology themes. First, the automobile and its supporting infrastructure clearly illustrate what is perhaps the most critical theme of industrial ecology: the need to view every activity and artifact within its systems context (an idea explored in more detail in Chapter 4). This does not mean that every Design for Environment (DFE) activity must be undertaken only if the entire system is affected: we should not try to link CO_2 cleaning of electronic automobile parts during manufacture, for example, with effects of highway bridge construction. It does mean, however, that activities undertaken at one level of the automotive technology system should not ignore relevant linkages with other levels of the system. Thus, as we shall see, much of DFE in the automotive sector is driven by needs defined at other, non-manufacturing, levels of the system—most obviously, the need to reduce CO_2 (global climate change), and NO_x and VOC (volatile organic carbon) emissions (tropospheric smog), while reducing energy consumption per vehicle kilometer.

Second, in Chapter 3 we focus on the integration of science and technology with another critical aspect of human society which makes efforts to reduce the environmental impacts of automobiles far more complex and problematic: culture. As we have already suggested, perhaps no other artifact is as culturally charged as the automobile. This factor in itself makes changing usage patterns to reduce environmental impacts daunting. With regard to economics, the bounded free market systems in countries such as the U.S. has resulted in the spontaneous emergence of a very efficient, albeit diffuse, post-consumer automobile recycling system. Unless well thought out, regulatory interference with this system in the name of environmental mitigation could disrupt it, ironically increasing, not decreasing, the lifecycle environmental impacts of the car.

Finally, this is a book focused most directly on the student or practicing engineer. We therefore provide the student with conceptual and actual tools, illustrated with reference to the automobile, to develop a capability to understand and apply DFE principles to complex manufactured articles such as automobiles. While the methods we present here are in their infancy, they still provide an increasingly rigorous basis by which the design and manufacturing engineer can integrate environment and technology in the products for which she or he is responsible. This skill is increasingly required, and sought after, by employers.

In the next chapter, the field of industrial ecology is briefly introduced. Then follows a chapter that describes a critical and unique aspect of the automobile for the industrial ecologist—its role as psychological and cultural icon. The introductory section concludes with a chapter discussing automotive technology as a system. Section II then provides a brief history of the evolution of the automobile and its supporting infrastructure. Section III is much more practical; it assumes that the focus is on the automobile as a specific product, and consists of chapters describing the DFE methodologies by which environmental impacts can be reduced. In Section IV, short, medium, and long-term projections for the evolution of the automotive sector and its infrastructure are provided.

SUGGESTED READING

Allenby, B. R. and D. J. Richards, eds., *The Greening of Industrial Ecosystems*, National Academy Press, Washington, DC, 1995.
DeCicco, J. and Ross, M., Improving automotive efficiency, *Scientific American, 271*(6), 52–57, 1994.
Hardin, G., The tragedy of the commons, *Science, 162*, 1243–1248, 1968.
MacKenzie, J. J., R.C. Dower, and D.D.T. Chen, *The Going Rate: What It Really Costs to Drive*, World Resources Institute, Washington, DC, 1992.

EXERCISES

1.1 Create a simple model of the automobile technology system, clearly identifying different levels of the system (for example, the engine system would be a different level than the automobile itself). Include the linkages that exist between the different levels (for example, "lightweighting" automobiles improves the efficiency of the engine in terms of distance traveled per unit of fuel). Save this model and elaborate on it as you read this text.

1.2 Make a list of all the different aspects of the automobile as an object, its use and functions, and the supporting infrastructures. For each aspect listed, determine whether it is primarily a result of technology and technological constraints or culture and individual psychology, and suggest short term and long term activities that could mitigate the resulting environmental impacts.

1.3 Make a list of the reasons that might lead you to purchase an automobile. For each reason, identify desirable related automotive technologies or characteristics, as well as any non-automotive alternatives that satisfy the same reason, whether they involve transport (e.g., a bicycle) or not (e.g., the Internet). Finally, discuss why each alternative meets, or fails to meet, the need you originally identified.

1.4 Express the following English unit quantities in the metric equivalents indicated (see Table B1 for conversion factors): (a) 30 mi/hr in km/hr; (b) 3 acres in hectares; (c) 3 acres in square meters; (d) 180 horsepower in watts; (e) 20 mi/gal in km/l.

1.5 Fig. 1.1 demonstrates that the 189 million automobiles in the U.S. in 1990 emitted about 30 million metric tons (30 Tg) of carbon monoxide while driving 3.4×10^{12} km. In 1970, the 108 million vehicles existing at that time traveled 1.8×10^{12} km, and their CO emissions per km were 96% higher than in 1990. How much CO did U.S. vehicles emit in 1970?

CHAPTER 2	# Industrial Ecology

"Industrial ecology" is intended to mean both the interaction of global industrial civilization with the natural environment and the aggregate of opportunities for individual industries to transform their relationships with the natural environment. It is intended to embrace all industrial activity . . . ; both production and consumption; and national economies at all levels of industrialization.

R.H. Socolow, *Industrial Ecology and Global Change*
Cambridge, U.K.: Cambridge Univ. Press, 1994.

2.1 THE MASTER EQUATION

Naive discussions of global environmental perturbations tend to assume global climate change is the major, or perhaps only, concern. This is simplistic. It is quite true that the exponential growth of emissions of critical chemical species known to force global environmental impacts closely parallels the Industrial Revolution; Figs. 2.1 and 2.2 demonstrate this pattern for CO_2, a combustion-related gas that forces global climate change.

Human environmental impacts, however, are not fully measured by any one perturbation, such as the increasing atmospheric concentrations of anthropogenic climate forcing gases, including CO_2, CH_4, H_2O, and perfluorocarbons. Rather, it is evident in the suite of regional and global impacts, including loss of biodiversity; increased global diffusion of toxics, both organic and inorganic; loss of arable soil; loss and degradation of fresh water supplies; stratospheric ozone depletion; global climate change forcing; and increased acidity of precipitation. In many cases, these impacts arise from the sheer magnitude of the effects that human global economic activity has on the stocks and fluxes of natural (i.e., pre-human) materials flows, from heavy metals (Table 2.1) to nitrogen (Fig. 2.3). While some

8

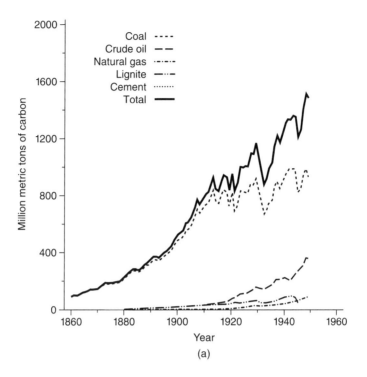

(a)

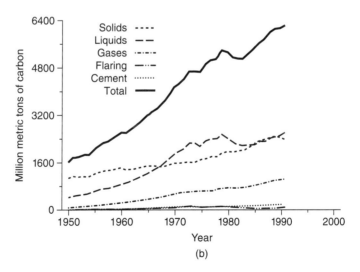

(b)

Figure 2.1   Global CO_2 emissions from fossil fuel burning, cement production, and gas flaring, 1860–1991 (T.A. Boden, D.P. Kaiser, R.J. Sepanski, and F.W. Stoss, *Trends '93, A Compendium of Data on Global Change*, Oak Ridge, TN: Oak Ridge National Lab. Rpt. ORNL/CDIAC-65, p. 502, 507, 1994.)

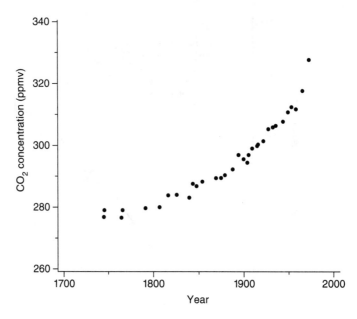

Figure 2.2   The increase in atmospheric carbon dioxide since 1700. (J.T. Houghton, G.J. Jenkins, and J.J. Ephraums, Eds., *Climate Change: The IPCC Scientific Assessment*, Cambridge, U.K.: Cambridge University Press, 1990.)

Table 2.1   Worldwide Atmospheric Emissions of Trace Metals (Gg/yr)

Element	Energy Production	Smelting, Refining	Manuf. Processes	Commercial, Waste Tmt., Transport	Total Anthro. Flux	Total Natural Flux
Antimony	1.3	1.5	—	0.7	3.5	2.6
Arsenic	2.2	12.4	2.0	2.3	19.0	12.0
Cadmium	0.8	5.4	0.6	0.8	7.6	1.4
Chromium	12.7	—	17.0	0.8	31.0	43.0
Copper	8.0	23.6	2.0	1.6	35.0	6.1
Lead	12.7	49.1	15.7	254.9	332.0	28.0
Manganese	12.1	3.2	14.7	8.3	38.0	12.0
Mercury	2.3	0.1	—	1.2	3.6	317.0
Nickel	42.0	4.8	4.5	0.4	52.0	2.5
Selenium	3.9	2.3	—	0.1	6.3	3.0
Thallium	1.1	—	4.0	—	5.1	29.0
Tin	3.3	1.1	—	0.8	5.1	10.0
Vanadium	84.0	0.1	0.7	1.2	86.0	28.0
Zinc	16.8	72.5	33.4	9.2	132.0	45.0

Source: J. Nriagu, Global metal pollution, *Environment, 32* (7), 7–32, 1990.

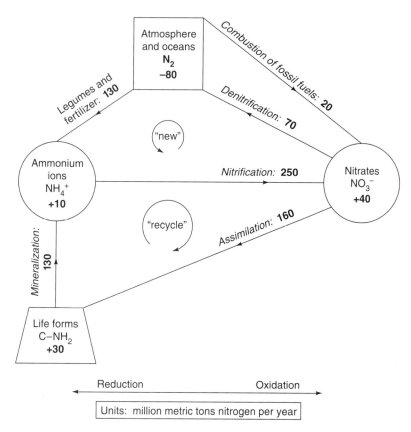

Figure 2.3   Human additions to the preindustrial global nitrogen cycle. Changes in reservoir stocks are shown by entries inside the boxes and sum to zero. (Schlesinger et al., *Industrial Ecology and Global Change*. Ed. R. Socolow et al., Cambridge, U.K.: Cambridge Univ. Press, 1995).

of these impacts arise from cultural or technology choices for which there are at least theoretical options, taken as a whole they represent a unique phenomenon in the history of the globe: the dominance of the human species on virtually every global and regional physical, chemical, and biological system.

Estimating the potential economic and social costs, and even the potential for a significantly reduced human carrying capacity of many regions (especially poor developing areas such as Saharan Africa) which could result from these perturbations is difficult. Nonetheless, available data, and the fact that many of these systems are quite complex and therefore quite likely to exhibit unpleasant discontinuities in their behavior, strongly indicates that, taken as a whole, the environmental impact of our species must not just be slowed, but reduced in absolute terms.

The fundamental stresses on global systems are perhaps best understood by considering the "master equation," which links environmental impact with population, the quality of life sought, and the technology with which that quality of life is provided:

$$\text{Environmental Impact} = \text{Population} \times \frac{\text{GDP}}{\text{Person}} \times \frac{\text{Environmental Impact}}{\text{Unit of GDP}} \qquad (2.1)$$

where GDP is a country's gross domestic product, a measure of national economic activity. Let us examine each term of the equation more closely.

2.1.1 Population

Human population growth is clearly an important contributing factor to increased anthropogenic environmental impact. It is generally recognized that the human population growth has exhibited explosive growth since the beginning of the Industrial Revolution, which, in conjunction with the concomitant revolution in agricultural technology and efficiency, in essence created unlimited resources for such a population explosion. It is intriguing that this pattern can be clearly seen, for example, in the population figures for England and Wales during the period 1650 to 1960 (Fig. 2.4), which clearly resembles the pattern of any species suddenly provided with unlimited opportunities for growth (Fig. 2.5).

What is less recognized, but is critical, is how closely globally sustainable levels of population are correlated with technological and cultural evolution. Thus, Fig. 2.6, based on work by Edward Deevey, Jr., shows that the three great jumps in human population have accompanied the initial development of tool use, the shift from hunter-gatherer cultures to agricultural cultures, and the Industrial Revolution. This pattern should not be read to imply specific links between given technologies and population levels. Human society is far more complex than that. Indeed, some experts, such as Joel Cohen in his authoritative book *How Many People Can the Earth Support?*, believe that no simple models adequately describe human population dynamics. Our current state of knowledge does suggest, how-

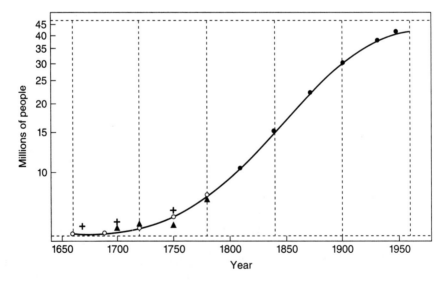

Figure 2.4   The population of England and Wales from 1650–1960. (A.P. Usher, *A History of Mechanical Inventions*, Revised ed., New York: Dover Pubs., 1954.)

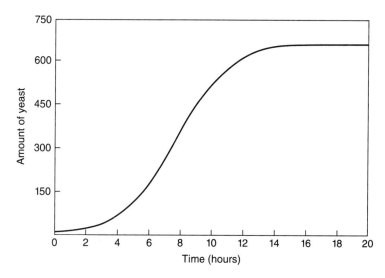

Figure 2.5   The growth of yeast in a culture medium of fixed size. (After E.P. Odum, *Fundamentals of Ecology*, Philadelphia: W.P. Saunders Co., p. 186, 1959.)

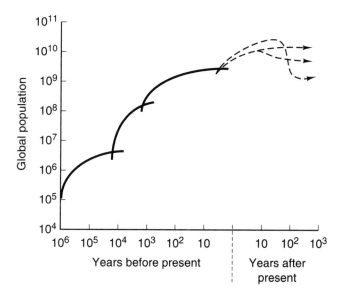

Figure 2.6   Population growth and stages of human cultural evolution. From left to right, the historical stages are tool use, the agricultural revolution, and the industrial revolution. The fourth (present and future) stage, shown in dashed lines, is that of ubiquitous technology and substantial environmental impact. Three possible scenarios for the fourth stage are pictured: one in which population stability is achieved by gradual, ordered approaches, one in which a reduced population stability occurs through a directed program of decreased use of technology, and one in which it is achieved by unmanaged growth followed by an unmanaged crash. (Based on E.S. Deevey, Jr., The human population, *Scientific American, 203*, no. 3, 194–206, 1960, and M.G. Wolman, The impact of man, *EOS-Trans. AGU, 71*, 1884–1886, 1990.)

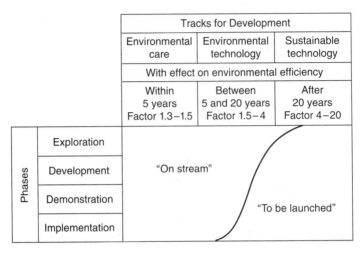

Figure 2.7   Phases for technology tracks having effects on environmental efficiency in The Netherlands. (Reproduced through the courtesy of J.L.A. Jensen.)

ever, that a reasonable response—perhaps, in the short term, the only reasonable response—to unsustainable environmental forcing is the rapid evolution of environmentally preferable technologies. Indeed, experts in The Netherlands, at present one of the most sophisticated countries in the world in terms of policies regarding sustainability, have focused on radical improvements in technology as a key to a sustainable global economy. Improvements in environmental efficiency by factors of 1.3 to 1.5 are seen as possible in the short term, with improvements by as much as 400 percent eventually possible with technology that is already developed (Fig. 2.7). Further substantial improvements will require exploration of new technology options and will take considerably longer to implement.

The human population continues to expand rapidly. Since 1970, human population has grown eightfold. It is now approximately 5.7 billion, and projections of a peak of 10 to 12 billion in the next century are not regarded by demographers as unrealistic. The timing and height of the eventual peak are difficult to predict with accuracy, and involve not only a number of cultural and political issues, but the fundamental, and unresolved, question of Earth's carrying capacity for our species given current technological systems. What is clearly apparent, however, is that the first righthand term in the master equation will be trending strongly upwards, likely by a factor of about two in the next half-century, unless some unforseen catastrophe causes a population collapse.

2.1.2   Per Capita Wealth

Like the first term, the second term of the master equation is trending strongly upwards. While wealth per capita varies significantly among countries and regions, and within political units may demonstrate substantial distributional disparities, the general trend is clearly positive, as Table 2.2 demonstrates. Citizens of developed countries are currently unwilling

Table 2.2   Growth of Real per Capita Income in More Developed and
Less Developed Countries, 1660–2000

Country Group	1960–1970	1970–1980	1980–1990	1990–2000
More Developed Countries	4.1	2.4	2.4	2.1
Sub-Saharan Africa	0.6	0.9	−0.9	0.3
East Asia	3.6	4.6	6.3	5.7
Latin America	2.5	3.1	−0.5	2.2
Eastern Europe	5.2	5.4	0.9	1.6
Less Developed Countries	3.9	3.7	2.2	3.6

The figures are average annual percentage changes. The 1990–2000 figures are estimated.

Source: The World Bank, *World Development Report 1992*, Oxford, UK: Oxford University Press, 1992.

to significantly modify their consumption patterns for environmental reasons, and citizens of developing countries are unwilling to limit their aspirations for a similar lifestyle.

It is seldom recognized how substantially economic activity has grown since the Industrial Revolution. Since 1700, the volume of goods traded internationally has increased some 800 times. In the last 100 years, the world's industrial production has increased more than 100–fold. In the early 1900's, production of synthetic organic chemicals was minimal; today, it is over 225 billion pounds per year in the United States alone. Moreover, many of the thousands of new materials introduced into commerce annually are not widely present in nature, so the eventual impacts of their use and dispersion, particularly on non-human species, may not be well characterized. Since 1900, the rate of global consumption of fossil fuel has increased by a factor of 50—much of it for transportation uses. What is important is not just the numbers themselves, but their magnitude and the relatively short historical time they represent.

At the present time, quality of life, which is what the second term really refers to, is almost universally defined in terms of ability to appropriate goods and materials. Owning things, especially articles such as automobiles that reflect status as well as possession, is an important goal for most of humanity. As Table 2.3 suggests, however, it is unlikely that the globe, already under significant stress from current levels of material consumption, would be able to support a world economy based on today's production and consumption levels. This point is reinforced for automobiles by Fig. 2.8, which demonstrates that the most populous developing countries have some two orders of magnitude fewer cars than does North America. Were this gap to begin closing rapidly, the strain on global material and energy resources would be substantial. Thus, at some point the link between perceived quality of life and material consumption must be weakened or broken. For now, however, the cultural changes necessary for this decoupling do not appear likely, and the trend of the second term of the master equation must be accepted as being strongly upward. Crude estimates suggest that an increase of a factor of perhaps 3–5 over the next half-century may well occur.

Table 2.3   Use of Selected Commodities in the Global Economy

Commodity	1990 U.S. Consumption	1990 World Production	Necessary Production for World per capita to Equal US per capita	Factor of Increase
Plastic	25.0	78.1	530.0	6.8
Synthetic fibers	3.9	13.2	82.7	6.3
Aluminum	5.3	17.8	111.5	6.3
Phosphate rock	4.4	15.7	93.3	5.9
Copper	2.2	8.8	46.0	5.2
Salt	40.6	202.3	860.7	4.3
Potash	5.5	28.3	115.5	4.1
Sand and gravel	24.8	133.1	525.3	4.0
Iron and steel	99.9	593.7	2117.9	3.6
Cement	81.3	1251.1	1723.1	1.4

All production and consumption figures are in Tg.

Source: U.S. Bureau of Mines

2.1.3   Environmental Impact per Unit of Production

The third term in the equation is the only one which, at least in the short term, offers any possibility of reducing the fundamental environmental impacts of human economic activity. Defined as the environmental impact per unit of per capita gross domestic product, it can be considered as the environmental impact necessary to provide a unit of quality of life to an individual. It represents the degree to which technology is available to meet human needs and aspirations without serious environmental consequences.

If in the next half-century the first term in the master equation will double and the second increase by a factor of 3–5, the third term will need to decrease by a factor of 6–10 just to maintain environmental impacts at today's levels. Because the present degree of many of those impacts is considered unsustainable, the technology term must eventually decrease by perhaps a factor of 20–50 to enable the possibility of sustainable development for society.

This result raises several important points. First, it encourages the broad definition of good technology as "the means by which society sustainably provides quality of life to its members." The advantage of such a definition is that it focuses on four critical dimensions of technology:

1. Technology must always be thought of as part of a system, which includes not just the artifact(s) in which technology is physically embodied, but also the psychological, cultural and economic systems within which that technology is embedded and used. Especially in the case of automobiles, which are as much expressions of individual and social psychology as physical objects, this is a critical consideration.

2. No technology should be considered good technology unless it is as environmentally preferable as possible within its economic and cultural context. The state-of-the-art

Figure 2.8   Rates of automobile ownership in countries around the world.

does not yet permit the definition of technologies, artifacts, or technological systems as "sustainable," but we clearly know enough to make environmentally preferable choices in many instances. The sum of such decisions will at least move the global economy in the right direction.

3. Technology should be thought of as the way in which a society provides quality of life to its members. In that way, continued development can be conceptually separated from ever-increasing levels of consumption of raw materials and energy.

4. Implicit in this approach to technology is the idea that it is not enough to simply develop a good technology. No technology will produce gains in economic and environmental efficiency until it is diffused throughout the economy. In many cases, technology development is relatively simple compared to the difficulties inherent in technology diffusion. Perhaps reflecting this, there is a far larger literature on technology development than technology diffusion, and no good theoretical basis for understanding technology diffusion has yet been developed.

The third term of the master equation also reinforces the critical message of Figs. 2.4, 2.5 and 2.6: unless we are implicitly or explicitly willing to accept unplanned reductions in human population, technology must be evolved that will continue to meet economic and cultural needs, yet do so in a more environmentally appropriate way than those that evolved during the Industrial Revolution (and reflect the lack of scarcities, and unconcern with any environmental impacts, which characterized that evolution).

Additionally, it is important to note that this focus on technology should not be confused with naive technological optimism, which, at its most extreme, holds that technological fixes will always mitigate any unforseen environmental impact or continued uncontrolled growth of population. Rather, it is a necessary realism to accept the world as it is—with a growing human population desiring higher levels of material wealth—and try to evolve human systems to support a rational path towards long-term sustainability under these constraining conditions. In the immediate future, only rapid technological evolution appears to be able to fill that role. While appropriate evolution of both culture and technology is necessary, it must be recognized that, absent a significant shock, it will be quite difficult, and take considerably longer, to achieve the social and cultural evolution that will be required to stabilize population growth and change consumption patterns which link material wealth and quality of life. In fact, the data do not definitively support even the guardedly optimistic assumption that, by immediate implementation of industrial ecology, economic, cultural, or demographic disruptions arising from environmental perturbations can be avoided. Rather, industrial ecology adopts as an operative assumption the possibility of a reasonably smooth transition to a stable carrying capacity; to assume otherwise runs the risk of either despair or extremism, and the creation of a self-fulfilling prophecy. It is not optimism, then, but realism which recognizes the critical role of industrial ecology.

Thus, the right side of the master equation indicates that achieving environmental and economic sustainability will be no simple task. The population and wealth per person terms are under strong upward pressure, with the latter perhaps even accelerating with the growth of the global economy. Whether and how the third term, which is basically a technology term, can compensate for this growth is the essence of industrial ecology.

2.2 DEFINITION OF INDUSTRIAL ECOLOGY

Industrial ecology can be simply defined:

> Industrial ecology is the study of the means by which humanity can deliberately and rationally approach and maintain a desirable carrying capacity, given continued economic, cultural, and technological evolution. The concept requires that an industrial system be viewed not in isolation from its surrounding systems, but in concert with them. It is a systems view in which one seeks to optimize the total materials cycle from virgin material, to finished material, to component, to product, to obsolete product, and to ultimate disposal. Factors to be optimized include resources, energy, and capital.

The words "deliberately" and "rationally" indicate that the intent of the multidisciplinary field of industrial ecology is to provide the technological and scientific basis for a considered path towards global sustainability, in contrast to unplanned, precipitous, and potentially quite costly and disastrous alternatives. "Desirable" indicates that, given the potential for different technologies, cultures and forms of economic organization, a number of sustainable states may exist. It then becomes a human responsibility to choose among them.

The differences between the industrial ecology approach and existing efforts to address environmental perturbations are summed up in Table 2.4. Of particular note is the fundamental shift in social and private firm treatment of environmental considerations: what was once overhead—that is, not linked with the primary purpose of economic activity—has now become strategic. Environmental management programs no longer focus simply on just one site, or single media, or specific substance impact, but broaden to view physical, economic and technology systems as a whole and over their entire life cycles. Similarly, activities are not driven just by one risk (e.g., human carcinogenesis), but reflect a need to move towards global sustainability, a far more difficult but far more meaningful endpoint. Compliance and remediation are important, but they are trivial compared to the challenge of achieving sustainability. An important initial step towards that goal is developing the necessary objective scientific and technological understanding, data and methodologies—and that is what the field of industrial ecology does.

A full consideration of industrial ecology would include the entire scope of economic activity, such as mining, agriculture, forestry, manufacturing, and consumer behavior. Pages 24–25 lay out the industrial ecology intellectual framework, which moves from the highest, most conceptual level—the vision of sustainable development - to application of the principles of industrial ecology as they are now understood in different sectors: DFE for manufacturing, IPM (integrated pest management) for agriculture, for example. While much current effort in industrial ecology is focused on manufactured articles and manufacturing, the field is far broader than that. Industrial ecology is, in fact, the "science and technology of sustainability."

In this volume, of course, the focus is on the automobile sector. Nonetheless, one of the attractions of this sector for the industrial ecologist is that the full scope of industrial ecology can, indeed, be glimpsed through this prism. The modern automobile includes materials produced in a number of ways: virgin and recycled metals, plastics from petro-

Table 2.4   Characteristics of Alternative Societal Approaches to the Interactions Between Technology and Environment

Characteristic	Remediation	Compliance	IE/DFE
Time Frame	Past	Present	Future
Activity Focus	Individual site, media, or substance	Individual site, media, or substance	Materials, products, services, and operations over life cycle
Endpoint	Reduce local anthropocentric risk	Reduce local anthropocentric risk	Global sustainability
Relation of environment to economic activity	Overhead	Overhead	Strategic and integral
Underlying conceptual model	Command and control intervention in simple systems	Command and control intervention in simple systems	Guided evolution of complex system
Policy leader	United States	Developed countries	Northern Europe, especially The Netherlands and Germany
Cost to firm of failure to perform	Liability, fine	Liability, fine	Loss of market access, uncompetitive products and services

chemicals, silicon in chips, and glass. It uses fuels which, increasingly, may come from biomass. Its tires can be eyesores or even mosquito breeding grounds when thrown away, but are increasingly recycled for energy or ground up for running tracks. It runs on roads and parks on lots in malls made from millions of tons of aggregate and built over thousands of hectares of previously green land and wetlands. And, finally, it is a psychological and cultural icon, raising significant issues of values and trade-offs, environmental, economic and otherwise. It is indeed a rich lode of understanding for the industrial ecologist.

2.2.1   Industrial Ecology Model Systems

The concept of industrial ecology can be illustrated by considering three different models of systems, as shown in Fig. 2.9. A Type I system is linear: virgin materials enter the system, are used only once, and then disposed of as waste. We will define "waste" in this book as material for which there is no further use within the system, as opposed to "residuals", which may have no further use within the generating process or firm, but can be reused within the system as a whole. It is a principle of industrial ecology that no economic

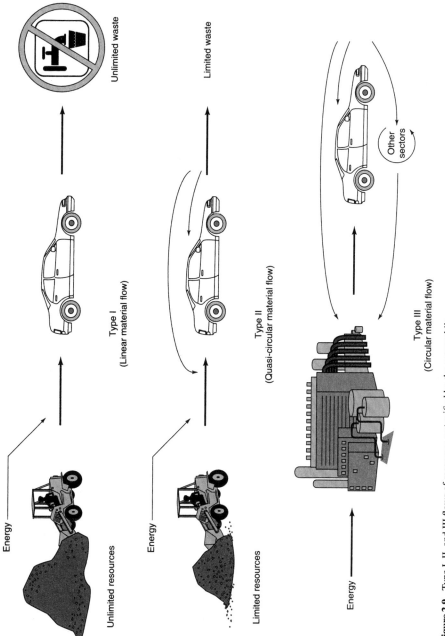

Figure 2.9 Type I, II, and III flows of resources as typified by the automobile.

activity should create waste, only residuals. Thus, an example of a Type I material flow would be an automobile that is abandoned by the side of the road at the end of its useful life—all waste. Note that a Type I system is the result not just of technology, but of social and economic factors as well: the car may be abandoned, for example, because the secondary metal markets are undeveloped, and thus unable to support the recycling of the metallic content of the car locally.

A more complex Type II system arises as scarcity (or growing population) makes the Type I system inadequate. Feedback and internal cycling loops develop, and flows of material into, and waste out of, the system diminish. Internal reuse of materials can become quite significant. For materials, the internal velocity of their use within the system may be high, but the velocity of materials through the system—from initial mining or processing to waste—is reduced. An example of a Type II system is the life cycle of most cars today in developed economies. The post-consumer car is first stripped of useful subassemblies, which are then recycled as used parts. The remaining hulk is shredded, and the steel, which is some 75 percent by weight of the automobile, is recovered for recycling. This is an internal material loop. The remaining plastic, glass, and miscellaneous materials and liquids, known as "fluff," are then landfilled. This stream constitutes waste, and should be redesigned or managed so as to become a residual.

The Type III system is one in which full cyclicity has been achieved. As with every product, Type III realization should be the goal of automobile design and management across the lifecycle, and, eventually, the goal of supporting infrastructure design as well. This vision will be extremely difficult to achieve, however, as some aspects of the automotive technological system are inherently "dissipative": that is, materials are degraded, dispersed, and lost to the economic system in the course of a single normal use. For example, tires leave dissipative residue on pavement, fuel is burned in combustion engines, and engine oil must be replaced periodically. In such instances, minimization of dissipative uses may be the best that can be hoped for, given current technologies. Nonetheless, even here there are opportunities: engine sensor systems now under development, for example, can determine when oil is physically dirty enough to require changing, and thus reduce the number of oil changes, now done on a routine and conservative schedule, perhaps substantially.

Note that all systems are energetically open, as solar energy is a constant input. Clearly, the constraint here is that energy use by the system as a whole must be sustainable; that is, the system must not rely on more energy than is constantly replenished by the sun (and, perhaps, nuclear fission and fusion should those technologies be developed to be environmentally and politically acceptable). It is also important to recognize that all cycles within the system tend to function on widely differing temporal and spatial scales, a behavior that greatly complicates analysis and understanding of the system.

While the essence of industrial ecology involves cycling of materials through the economy, thus supporting more economic activity with less material throughput, it would be far too simplistic to conclude that recycling is the answer to everything. In the case of lightweight hydrocarbon plastics, for example, localized energy recovery may be the most environmentally appropriate course of action. Moreover, the choice between what elements of a product to refurbish as is, to upgrade and reintroduce into commerce, or to recycle at the materials level is not trivial in many cases. For example, some have urged that the life-time of automobiles be extended to avoid premature disposal of their constituent materials. This position is questionable given data indicating that it is predominantly older cars which generate the most pollution on a per unit basis. Rather, a more sophisticated approach might be to design a car with modular subsystems, so that, for example, upgraded engine systems can be put in older chassis and the metal content of the used engine recovered.

2.2.2   Perspectives on Industrial Ecology

Taking another approach, Robert Socolow identifies six perspectives provided by industrial ecology. Industrial ecology focuses on long-term habitability rather than the short term issues characterizing current approaches; on mitigating disruptions to fundamental life-supporting cycles rather than just responding to obvious localized perturbations. Similarly, industrial ecology is characterized by global scope rather than transitory issues which are local in time and space. Industrial ecology seeks to identify, and avoid, instances where human activity overwhelms natural systems. The field is also fundamentally concerned with maintaining the resilience of human and natural systems and the logical corollary, identifying and protecting more vulnerable systems. Industrial ecology is also character-ized by a focus on mass-flow analysis, as the physical basis of economic activity and con-comitant environmental impacts are quantified. Finally, industrial ecology recognizes both in name and in practice the centrality of economic production agents such as farms and industrial firms. Understanding and modifying their behavior to be more environmentally appropriate, rather than blame or adversarial posturing, is seen as more productive. This trend is reinforced as attention is shifted from regulation of production, to management of product across its lifecycle.

SUGGESTED READING

Graedel, T. E. and B. R. Allenby, *Industrial Ecology*, Prentice-Hall, Inc., Englewood Cliffs, NJ, 1995.
Socolow, R., C. Andrews, F. Berkhout and V. Thomas, eds., *Industrial Ecology and Global Change*, Cambridge University Press, Cambridge, 1995.
World Commission on Environment and Development (the Brundtland Commission), *Our Common Future*, Cambridge University Press, Cambridge, 1987.

EXERCISES

2.1   A car is a combination of many subassemblies, components, and materials, some of which exhibit Type I behavior, some of which exhibit Type II behavior, and some of which (potentially) exhibit Type III behavior.
   a. Identify 5 major components and 5 major materials and determine whether their behavior is Type I, Type II, or Type III.

b. Identify ways in which increased cyclization of each selected component and material can be achieved. Do not limit yourself to technological solutions: in some cases, for example, the key to increased cyclization may be economic or political.

c. In each case, determine whether the increased cyclization is environmentally preferable to responsible disposal. What additional data would help you defend your conclusions?

2.2 Industrial ecology is sometimes referred to as "the science of sustainability." Like any such catch phrase, this is both over-simplistic and true. Discuss both aspects, using the automobile as an example.

2.3 The last term of the "master equation" is written as "environmental impact per unit of GDP". What would be the advantages of rephrasing it as "environmental impact per unit quality of life"? How would you measure quality of life?

2.4 The consumptive water use chargeable to the average American is about 1400 kg/day. How large a volume is this? How large a volume is a year's water use?

2.5 It is estimated that the materials use, including construction materials and fossil fuels, chargeable to the average American is about 50 kg/day. Japan's figure is similar, while people in most other countries use much less. What level of materials use does this imply if all the world's population used materials at this rate? What does it imply for the year 2050, when today's 5.7 billion people may have increased to 12 billion? Assuming an average density for the materials of 2 g cm^{-3}, what is the daily volume of the materials that will be extracted from their reservoirs?

THE INDUSTRIAL ECOLOGY INTELLECTUAL FRAMEWORK

Without an intellectual framework within which to operate, it is difficult to avoid confusion, especially when one is dealing with complex systems (see Chapter 4). Accordingly, the framework on the following page, though simple, may be of much value to the reader. "Sustainable development" was originally defined by the Brundtland Commission as "development that meets the needs of the present without compromising the ability of future generations to meet their own needs." This is a worthy and necessary vision, but is inherently ambiguous. It is also impossible to operationalize: although some talk of "sustainable firms" or "sustainable products," neither metrics nor a firm understanding of what sustainability really is underlie such concepts. "Industrial ecology," the "science and technology of sustainability," is the objective, multidisciplinary study of industrial and economic systems and their linkages with fundamental natural systems, providing the theoretical scientific basis upon which understanding and reasoned improvement of current practices can be based. The "industrial ecology infrastructure" encompasses all the necessary support systems, including legal and economic structures, methodologies and tools, and data and information resources, society must provide to individuals, firms and other organizations to support their implementation of industrial ecology. The next level of the system is where current applications of industrial ecology principles are beginning. It includes activities in a number of sectors: in manufacturing, for example, development and application of Design for Environment (DFE) methodologies are advancing fairly rapidly among electronics and automotive manufacturing firms. DFE development and deployment is paralleled in agriculture, for example, by similar activities involving IPM (Integrated Pest Management). DFE also includes research programs, in many cases already under way, targeted at developing important components of the industrial ecology infrastructure.

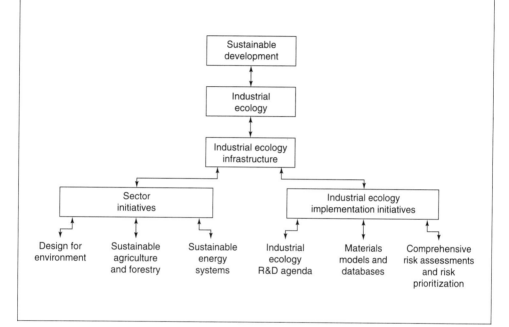

Culture, Psychology, and the Automobile

"Glorious, stirring sight!" murmured Toad, never offering to move. "The poetry of motion! The real way to travel! The only way to travel! Here to-day—in next week to-morrow! Villages skipped, towns and cities jumped—always somebody else's horizon! O bliss! O poop-poop! O my! O my! . . ."

"And to think I never knew!" went on the Toad in a dreamy monotone. "All those wasted years that lie behind me, I never knew, never even dreamt! But now—but now that I know, now that I truly realise! O what a flowery track lies spread before me, henceforth! What dust-clouds shall spring up behind me as I speed on my reckless way! What carts I shall fling carelessly into the ditch in the wake of my magnificent onset!"

> Toad in K. Grahame, *The Wind in the Willows*,
> after his cart is run off the road by an automobile

I think that cars today are almost the exact equivalent of the great Gothic cathedrals: I mean the supreme creation of an era, conceived with passion by unknown artists, and consumed in image if not in usage by a whole population which appropriates them as a purely magical object.

> B. Barthes, *Mythologies*, 1957

3.1 THE AUTOMOBILE AS CULTURAL ICON

A fundamental mistake made by many scientists and technologists studying industrial ecology is to assume that the major issues and problems are technical in nature. This is, in fact, infrequently the case. Rather, while elements of new technological systems or scientific data on, for example, the lifecycle impact of specific materials, may certainly need to be developed, it is often non-technical issues that prove most unfathomable to industry in general and the industrial ecologist in particular. Nothing illustrates the point better than the most emotionally-charged artifact the Industrial Revolution has ever produced: the automobile.

What do automobiles represent in the popular mind? From the very beginning, they have only tangentially represented the function that they purportedly perform, transportation. This truth is perhaps best captured in the history of automobile advertising, both a motivator and a mirror of the passions which automobiles arouse around the world.

Initially, from the 1890's until the early 1900's, advertising focused on the advantages of automobiles over horses, reflecting the need to shift consumers from the older, entrenched fundamental technology system. As is frequently the case, these advertisements tended to be phrased in terms of the more familiar technology they sought to displace. As a Winton car advertisement of 1905 put it: "The Winton car is as sensitive as a well-trained horse and ten times as reliable . . . The horse might get scared at the sight of a motor car, but the Winton can't get scared nor get tired."

This image changed dramatically with the advertisements created by Ed Jordan, manufacturer of a rather pedestrian car known as the Playboy, who, complaining that existing automobile advertisements were all done by engineers, pioneered between 1918 and 1926 a new, emotional style of advertisement (a contemporary British example is provided in the box on page 28).

The evolution in advertising, and rapid technological and business change in the automobile sector, went hand-in-hand. Significantly, by about 1925 the engineering approach to building cars exemplified by Ford—unchanging, universal styles—was being replaced by General Motors's Sloanism—rapid obsolescence supported by annual model changes, driven by sophisticated psychological advertising. This evolution was clearly motivated by the fact that the market for transportation alone—exemplified by the Ford—was becoming saturated. Sloan realized that continued profit growth, his primary goal, could only be achieved by a fundamental shift: sell a dream, make cosmetic changes annually, and make consumers believe they had to keep pace. As he said, "create demand for the new value and, so to speak, create a certain amount of dissatisfaction with past models as compared with the new one." One no longer bought a ton and a half of engineered machine which simply replaced a horse. One bought a lifestyle—indeed, one defined a new lifestyle through the choice of a car.

Thus, through skill on the part of the advertisers and as a logical extension of the machine itself, early on the automobile began to represent, in a very fundamental way, personal freedom:

> Dear Sir: While I still have got breath in my lungs I will tell you what a dandy car you make. I have drove Fords exclusively when I could get away with one. For sustained speed and freedom from trouble the Ford has got every other car skinned and even if my business hasn't [sic] been strictly legal it don't hurt enything [sic] to tell you what a fine car you got in the V8—
>
> Yours Truly,
> Clyde Champion Barrow

(presumably unsolicited promotional letter dated 10 April 1934 to Henry Ford from Clyde Barrow of the outlaw team of Bonnie and Clyde)

That this attribute of automobiles, which some associate particularly with the United States, is, in fact, universal is demonstrated by the British advertisement for Jowett automobiles in the box below.

THIS FREEDOM

For a thousand years we Englishmen have fought for our freedom. Our home is our castle; the highway winds unbarred. Every river is musical with memories; every green field, every beacon hill, is rich with the dust of those who fought for "This Freedom." Have you spied the purple iris blossoming along the river bank? Have you glimpsed a bit of heaven whilst "picnicing" by the scented pinewood? The wind's on the heath, brother; the highway is calling; there's laughter and deep breath and zestful life over there on the hills. Freedom is waiting you at the bang of your front door. "Is it possible," you ask, "that this freedom can be mine?" Once you ask that question a Jowett [automobile manufactured by Jowett Car Company, Yorkshire, Great Britain] is beginning to drive up to your door. For sixteen years the Jowett Car has been bestowing "This Freedom" on grateful folks, at a price you can easily afford. Freedom! At less than you pay for the humdrum, crowded railway train. Read on, and you will ride on—Brother of the Broad High Way!

THIS FREEDOM IS YOURS

(1920's British advertisement, reproduced in Pettifer and Turner, 1984)

Also from the beginning, the freedom was not only spatial—the open road—but psychological, especially, and sometimes blatantly, sexual. The 1941 Nash sedan, for example, contained a full six-foot double bed. As cars grew more powerful, the association between appearance, horsepower, and (adolescent) masculinity became more obvious (and in many cases more absurd). Consider the ad copy for the 1968 Toronado: "Toronado. The all-car car for the all-man man. The line of demarcation is drawn. Men on one side. Boys on the other. Cars fall into place." The early days of rock-and-roll—with songs such as "Little Deuce Coupe," "409," "Little Old Lady from Pasadena," "Little Cobra," "GTO," "Shut-down," and "Dead Man's Curve" - paid tribute to, and reinforced, this relationship.

Despite the prevailing image of cars as masculine sexual symbols, however, the more subtle and fundamental impact may have been on the sexual freedom of women: as Pettifer and Turner note, "in a social system that oppressed women in particular, [the automobile's] emancipating influence was profound, not least in the sexual field . . . according to some psychologists, women responded instantly to the attractions of the car because, as members of the relatively powerless sex, it gave them great satisfaction finally to have power under their control." The following advertising copy for the 1965 Mustang illustrates this psychology nicely:

Life was just one diaper after another until Sarah got her new Mustang. Somehow Mustang's sensationally sophisticated looks, its standard-equipment luxuries (bucket seats, full carpeting, vinyl interior, chiffon-smooth, floor-mounted transmission) made everyday cares fade far, far into the background. Suddenly there was a new gleam in her husband's eye. (For the car? for Sarah? both?) Now Sarah knows for sure: Mustangers have more fun!

The Mustang in particular seems to have had this ambiance: consider the James Brown rock-and-roll lament in "Mustang Sally" that, upon buying his girl a Mustang, she immediately takes to cruising—without him. He has, in other words, unwittingly purchased her personal freedom.

3.2 OWNERSHIP, USE, AND POPULATION DISPERSION

The freedom of the automobile translates as well into freedom in space and time. No longer is it out of the question to go to a concert 50 km away and return home the same evening. No longer is it impossible to take an evening art course at the university 30 km distant. No longer is it beyond reality to plan to organize a choir whose members may come each week to rehearsals from cities comprising a circle of 80 km radius. How about driving from southern England to northern Scotland for the weekend?

Statistics reflect in many ways the evolution of society that the automobile has had such a major role in bringing about. The simplest measure is the increase over time in vehicle distance traveled each year. For the U.S., this is reflected in a 440% growth between 1950 and 1992, during a period when population increased only 68% (Fig. 3.1). Furthermore, this growth has increased for drivers of all ages, and for both men and women (Fig. 3.2). Much of this driving is not for business or shopping, but for "leisure": vacations, a night at the football game, the daughter's ballet lesson (Fig. 3.3).

As driving habits have changed, so have drivers. Figure 3.4 shows German car ownership data for 1982 and 1990. Among the interesting features of this figure are the following: (1) Males are two to three times as likely as females to own a vehicle; (2) Car ownership increased about 10% for males and about 50% for females from 1982 to 1990; (3) Car ownership increased from 1982 to 1990 for all age levels from 25 to 85 years for both males and females; and (4) Elderly people are more likely each year to maintain vehicle ownership than was the case with their predecessors. These trends, which doubtless are reflected in other countries and regions of the world, indicate that car ownership (and presumably operation) is rapidly becoming a routine feature of daily life throughout much of the world, only weakly dependent on such characteristics as age, gender, or income level.

It is too simplistic, however, to assume that all societies respond identically to the opportunities available for public transportation; population density, tradition, the existing structure of cities and towns, and individual incomes vary too much for that. Table 3.1 provides some vehicle statistics for seven countries with moderate to high technological development. The variations are striking. Per thousand people, all the countries have 300–600 automobiles except Mexico, which has about 80. Motorcycle/moped ownership is very low except in Japan, where it is about 40% of automobile ownership. The number of buses per

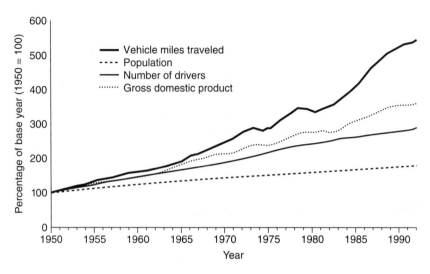

Figure 3.1   The historical growth in U.S. population, gross domestic product, and annual average vehicle distance traveled. (After G. Keoleian, K. Kar, M. Manion, D. Menerey, and J. Bulkley, *Industrial Ecology and the Automobile*, National Pollution Prevention Center Report, University of Michigan, Ann Arbor, Jan. 15, 1996.)

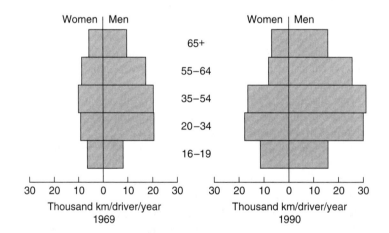

Figure 3.2   Vehicle miles traveled per year as a function of driver age, for U.S. drivers in 1969 and 1990. (Adapted from L. Schipper, Life-styles and the environment: The case of energy, *Daedalus, 125* (3), 113–138, 1996.)

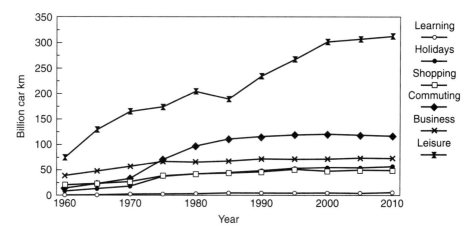

Figure 3.3   Distance traveled by automobiles in Germany subdivided by type of travel. (Adapted from *Car 2000*, Motor Industry Research Association, 1993).

thousand people varies only by a factor of two, and in no discernable pattern. Canada has the most trucks per capita, but the U.S. and Japan are nearly as high; the European countries and Mexico have far less.

The distance traveled per vehicle also shows wide variations among countries. The United Kingdom driver, for example, travels in his car nearly as far in the average year as does the Canadian or American, and farther than the German or the Swede (Table 3.2). Canadian and Mexican motorcyclists go similar distances, but far less than the German. German buses travel farther each year than those elsewhere. Truck distances are much more uniform, with only Mexican trucks having very much lower annual use than those elsewhere.

Vehicle ownership and distance traveled reflect the degree to which the automobile has resulted in demographic patterns that have deified its importance, particularly in the United States, the largest automobile market. The psychological personal freedom embodied in the technology has been translated into the geographic freedom to live in suburbia, then in exurbia. Jobs have moved from central city locations to suburban office parks; shopping from the city to outlying malls, originally massive centers for material acquisition, but now for entertainment and social interaction; road systems from hub-and-spoke systems linking urban centers to webs of pavement providing easy access to virtually all areas, at least in developed countries.

Moreover, living patterns have changed as well. Two income families require two commuting vehicles, and weekends full of organized regional sports for the children and shopping errands that can't be done during the week also create transportation demands for which a single automobile per family is inadequate. In some countries such as the United States, therefore, a culture has arisen where even going from a two-car to a one-car family is, for many consumers, essentially unthinkable.

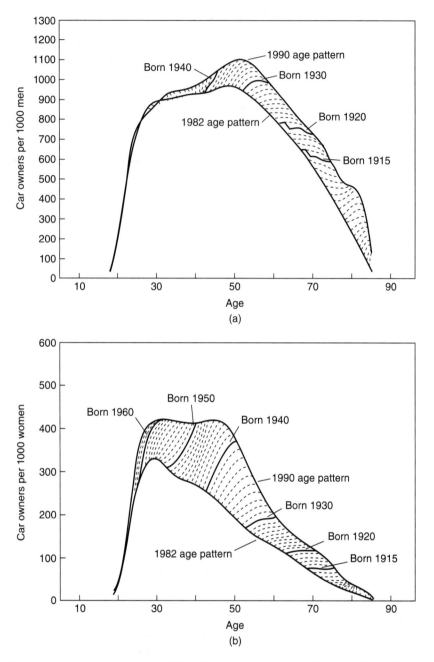

Figure 3.4   German car ownership rates for 1982, 1990, and selected cohorts: (Top panel) Males, (Bottom panel) Females (after T. Büttner and A. Grübler, The birth of a "green" generation? Generational dynamics of resource consumption patterns, Working Paper WP-94–79, Intl. Inst. for Applied Systems Analysis, Laxenburg, Austria, 1994).

Table 3.1   Number of Vehicles in Selected Countries

	Japan	Germany	Sweden	UK	Canada	Mexico	U.S.
Autos							
(millions)	40.3	24.4	3.6	20.3	13.3	7.5	146.3
per 1K people	327	424	409	363	480	83	567
Mopeds/cycles							
(millions)	16.4	2.4	0.11	0.71	0.34	0.26	4.0
per 1K people	132	29	13	13	12	3	15
Buses							
(thousands)	248	86	14	107	64	106	654
per 1K people	2.0	1.1	1.6	1.9	2.3	1.2	2.5
Trucks							
(millions)	22.3	3.8	0.3	3.9	6.9	3.5	47.1
per 1K people	179	47	35	69	249	39	183

U.S. truck data is 1990; other truck data are for 1994. Other data for Canada, Germany, Mexico, and U.K. are for 1992; for Japan, Sweden, and U.S. are for 1993.

Source: U.S. Federal Highway Administration, *Highway Statistics*, 1994.

Table 3.2   Vehicle Travel Distances in Selected Countries (km)

	Japan	Germany	Sweden	UK	Canada	Mexico	U.S.
All autos/yr	4.0(11)	4.6(11)	4.3(10)	3.3(11)	2.2(11)	3.6(10)	2.6(12)
per vehicle/yr	9.9(3)	1.2(4)	1.2(4)	1.6(4)	1.7(4)	4.8(3)	1.8(4)
All Mopeds/cycles/yr	ND	1.1(10)	ND	4.5(9)	1.2(9)	1.1(9)	1.6(10)
per vehicle/yr	ND	4.7(3)	ND	6.3(3)	3.5(3)	4.1(3)	4.0(3)
All Buses/yr	7.0(6)	4.4(6)	ND	3.3(6)	2.0(6)	5.3(5)	9.8(6)
per vehicle/yr	2.8(4)	5.1(4)	ND	4.3(4)	3.2(4)	5.0(3)	1.5(4)
All Trucks/yr	2.6(11)	4.1(10)	ND	6.4(10)	7.7(10)	1.6(10)	1.1(12)
per vehicle/yr	1.2(4)	2.6(4)	ND	1.6(4)	1.1(4)	4.4(3)	2.2(4)

Data are for varying years from 1990–1993. 4.0(11) indicates 4.0×10^{11}. ND = no data.

Source: U.S. Federal Highway Administration, *Highway Statistics*, 1994.

Thus, the association of the automobile with individual freedom is both psychological and deeply embedded in the physical patterns of modern society. Trapped in a traffic jam or driven frantic by weekend driving errands, the modern utilizers of automotive technology may perhaps be forgiven the feeling that they have unwittingly become party to a Faustian bargain (see Fig. 3.5).

Just how Faustian this bargain may be is indicated by the way that the automobile has thus embedded demand in the psychological and cultural patterns of many nations. By making possible two-career commuting families, dispersed community activities, and the separation of work from shopping from residence, the automobile has created a "freedom" that requires people to be chained to automobiles moving slowly on crowded highways for as much as several hours a day. Mass transit is incapable of serving such demand patterns, because of the dispersal of population that the automobile has facilitated. Moreover, this approach to life and society cannot be easily or quickly changed, as much of the buildings and the infrastructure that support it in such advanced automotive economies is already in place, creating an enormous potential cost to any change.

3.3 IMPLICATIONS FOR INDUSTRIAL ECOLOGY

Unlike elsewhere in this book, we have provided in this section a number of excerpts from a variety of sources about automobiles. They are interesting, even amusing, in themselves. Taken as a whole, however, and combined with the feelings about automobiles which many readers no doubt already bring to this book, they demonstrate beyond any doubt the truth that the automobile is not, either for existing cultures or individuals wherever they may live, simply a technological artifact. Whether this is desirable is immaterial: it is simply an objective fact that cars are psychologically charged icons. Any attempt to understand the industrial ecology of automobiles and their supporting technological systems—e.g., petroleum drilling, refining and distribution; road construction and concomitant aggregate and asphalt production—is incomplete unless all these dimensions are considered.

The automobile thus poses a number of challenging questions for the industrial ecologist. A fundamental one is the extent to which, or even whether, technological innovation can compensate for continuation of strongly embedded but unsustainable consumer preferences and behavior. Can increasingly more efficient cars really reduce the environmental impacts of automobiles taken as a whole, or does this represent an inherently unworkable attempt to substitute a technological solution for a dysfunctional cultural phenomenon (e.g., adoption of a technology with an inherently high social cost as an important source of perceived quality-of-life)? Alternatively, is it possible to use rapid technological innovation to "buy time," as the complex and difficult underlying cultural and psychological relationships between people and automobiles are redefined over the longer term? How rapidly can the automobile sector be evolved such that it contributes to global sustainability, given the technological, cultural, and economic constraints, and how do these constraints interact with each other?

The industrial ecology study of the automobile sector also raises interesting issues from an epistemological viewpoint. How, for example, can the tangled scientific, technical,

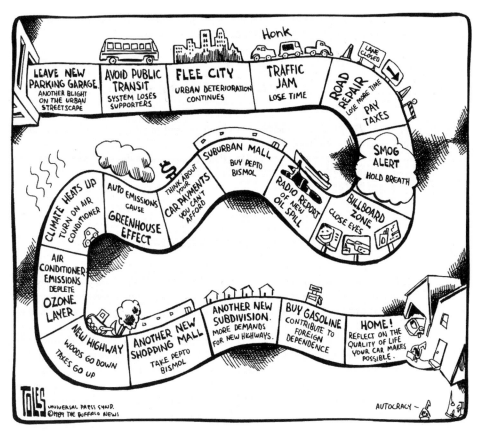

Figure 3.5   An editorial cartoonist looks at the mobile society. (Reprinted with permission of Universal Press Syndicate; copyright 1989 by *The Buffalo News*. All rights reserved.)

economic, cultural and psychological aspects of the automobile's role in society be separated for study without losing the characteristics of the actual status of the automobile which make it such a difficult problem in terms of reducing environmental impact? Some reductionism is obviously required, and yet the resultant understanding must be integrated so as to produce realistic, appropriate policies: how is this best accomplished? And what discipline, if not industrial ecology, offers the intellectual tools to do so? In all this, of course, it is necessary to maintain an objective, not normative, approach to the automobile culture: the industrial ecologist should seek to demonstrate what is so, in order to provide the knowledge for the political system to determine what should be so.

In this important sense, then, the automobile as artifact, as technological system, as industrial sector, and as a personally and culturally defining technology is an important study for the student of industrial ecology. Few other products of the Industrial Revolution illustrate so effectively the multidimensional nature of artifacts in general, and the need to consider them within the economic and cultural context of their actual use. And none raise so precisely, and so fundamentally, the potential conflicts between materialism and its contribution to perceived quality-of-life, and environmental impact of those choices.

SUGGESTED READING

Boyne, W. J., *Power Behind the Wheel: Creativity and the Evolution of the Automobile*, Stewart, Tabori & Chang, New York, 1988.

Grübler, A., The transportation sector: Growing demand and emissions, *Pacific and Asian Journal of Energy*, (2), 179–199, 1993.

Pettifer, J. and N. Turner, *Automania*, Little, Brown and Company, Boston, 1984.

EXERCISES

3.1 List a few reasons why you or people you know well purchased the car that they did. How many pertain primarily to "technical" aspects of automobiles (e.g., safety), and how many to psychological aspects (e.g., demonstration of wealth)? Which were most important in the buying decision?

3.2 List five common models of automobile and rank each one against the reasons for purchase that you identified in exercise 3.1. Which are most desirable and least desirable, and why? To what extent does the advertising you are aware of for these models focus on those characteristics?

3.3 Select a current issue of an automobile magazine and a personal computer magazine. Scan each for advertisements of automobile or PC models. In what ways is the language and imagery similar? How does it differ? Do any of the advertisements use language derived from one technology to describe the other (e.g., computer advertisements using the term "horsepower")?

CHAPTER 4

The Automotive Technology System

"When we try to pick out anything by itself, we find it hitched to everything else in the Universe."

John Muir, American naturalist

4.1 CONCEPTS OF SYSTEMS

Perhaps the most important operational feature of the industrial ecology approach is a focus not solely on the product itself, but also on its related systems and their behavior. Identifying the appropriate system and properly relating it to its technological context is frequently a critical step in any successful industrial ecology assessment. Accordingly, it is useful to briefly discuss technological systems in general, with particular attention to the difference between the simple and the complex.

A system may be thought of as a group of interacting, interdependent parts linked together by exchanges of energy, matter, and/or information. Defining a system is almost always somewhat arbitrary, requiring the analyst to identify those exchanges that are relevant to the purpose of the definition, to understand the linkages between the system as defined, and to determine the external context within which the system is embedded. For example, one would not want to try to define the environmental aspects of a technology within a system defined as the solar system—the system is too large and many of the linkages to the technology of interest are too indirect. Rather, one might choose a limited system—for example, what are the environmental effects in a province or region of the inputs, outputs, and processes related to a particular technology, and can they be reduced? Even here, data uncertainties and complexities may make it appropriate to identify and evaluate only the major impacts involved. For example, the life-cycle assessment (LCA) methodology developed by the Society of Environmental Toxicologists and Chemists (SETAC) is quite thorough, but has proven in practice to be too complex and expensive for most industrial uses. Accordingly, it has been replaced in a number of corporations with streamlined life cycle assessments (SLCAs) that attempt to identify the major life-cycle environmental issues in a much more efficient and rapid way.

4.1.1   Simple and Complex Systems

The most important distinction between classes of systems for the industrial ecologist is that between simple and complex. In many cases, the management approaches, regulatory structure, and analytical techniques traditionally applied to issues involving technology, economic efficiency, and environmental efficiency explicitly or implicitly assume the system involved is simple. In virtually all cases, however, the systems involved—whether they are economic, environmental, or technological—are complex. The terms are used here not in the sense of classical physics (e.g., the "simple" harmonic oscillator), but of the dictionary definitions, "complex" meaning "consisting of interconnected parts so as to make the whole difficult to understand"; "simple" is thus "not combined or compound". Some of the implications of this distinction are:

1. Simple systems tend to behave in a linear fashion: the output of the system is linearly related to the input. Complex systems, on the other hand, are characterized by strong interactions among the parts and by non-linear responses. For example, a salt marsh may be relatively resistant to chemical pollution until a threshhold is passed, after which even a small increment of additional insult will cause it to suddenly degrade precipitously. It is a complex system.

2. Simple systems can generally be readily evaluated in terms of cause and effect: an action is easily traceable through the system to its predictable effect. With complex systems, on the other hand, one finds feedback loops that often make the linkage between cause and effect quite difficult to establish.

3. Simple systems, unlike complex systems, are not characterized by significant time and space discontinuities, threshholds, and limits. For example, a problem for most people in comprehending the issues surrounding global climate change is that there are large time lags between the activity creating the forcing function, such as driving automobiles or using electricity, and the resulting shifts in global climate patterns. Compounding the time lags are spatial lags—the links between local activities, such as driving cars, and the global effects that may primarily be evident in distant regions (e.g., coastal flooding in Bangladesh) further attenuate comprehension. Because of such discontinuities, thresholds, and limits, most people do not intuitively comprehend the behavior of complex systems.

4. Simple systems tend to be characterized by a stable and known equilibrium point to which they return in a predictable fashion if perturbed. Many complex systems, on the other hand, operate far from equilibrium in a state of constant adaptation to changing conditions. Complex systems often evolve, simple systems generally stay more or less as they are.

5. With simple systems, change is assumed to be an additive function of subsystem characteristics: if you combine three quarters, you get $0.75, no more and no less. With complex systems, on the other hand, emergent behaviors may not be additive: one cannot predict the characteristics of an anthill by summing up the observed behavior of individual ants.

It is worth noting that recognizing the difference between simple and complex systems is not a new achievement. More than 2000 years ago, the proto-scientific Taoists in China made some insightful and provocative comments on the nature of such systems, quoted in Box 1.

SYSTEMS THINKING
(CA. 2200 BP)

To know that one does not know—that is high wisdom. The fault of those who make mistakes is that they think they know when they do not know. In many cases phenomena seem to be of one sort (alike) when they are really of quite different sorts. This has caused the fall of many states and the loss of many lives . . .

Lacquer is liquid, water is also liquid, but when you mix the two things together, you get a solid. Thus if you moisten lacquer it will become dry. Copper is soft, tin is soft, but if you mix both metals together they become hard. If you heat them they will again become liquid. Thus if you wet one thing it becomes dry and solid; if you heat a (hard) thing it becomes liquid. Thus one may see that you cannot deduce the properties of a thing merely by knowing the properties of the classes (of its components).

A small square is of the same class as a big square. A little horse is of the same class as a big one. But little knowledge is not of the same class as great knowledge. In the state of Lu there was a man called Kungsun Cho who said he could raise the dead. When they asked him how, he replied, "I can heal hemiplegia (apoplexy). If I gave a double dose of the same drug, I could therefore raise the dead." But among things there are some which can have small-scale effects, but not large-scale ones, and other things which can perform the half but not the whole.

By scholars gathered by Lu Pu-Wei, Lu Shih Chhun Chhiu (Master Lu's Spring and Autumn Annals, Taoist compendium of natural philosophy) 2240 BP, Tr. by R. Wilhelm, Fruhling u. Herbst d. Lu Bu-We, Diederichs, Jena, 1921, quoted in J. Needham, Science and Civilization in China, Cambridge University Press, Cambridge, 1991.

4.1.2 Evolution of Systems

Perhaps the most critical feature of complex systems is their evolution in response to changes in internal and external properties and conditions. This evolution occurs as the result of three linked mechanisms:

1. There must be some means of information storage and transmission;

2. There must be some mechanism for generating new alternatives for systems states (in biology, this mechanism is mutation; in economic systems, innovation); and,

3. There must be some selection from among alternatives, based on the performance of the alternatives in light of evolving internal states and external system boundary conditions.

In complex systems, as opposed to simple systems, this evolution is path dependent: that is, history matters. In the discussion of industrial ecology in Chapter 2, for example, we noted that we cannot ignore the presence on Earth today of 5.7 billion people in choosing future policy alternatives. That is another way of saying that the relationship between technology and population is a complex, co-evolved system, with an arrow of time.

Moreover, it is important to recognize that complex systems are "messy," in the sense that their evolutionary track can approach any of several stable or metastable states. This characteristic is reflected in the definition of industrial ecology that emphasizes a "desirable" carrying capacity: desirable implies choice, but if there is only one possible equilibrium point, there can be no choice involved. The study of industrial ecology supports, in fact, the ethical principle that our species can choose among alternative futures; whether we choose to do so rationally or not is a separate question. An obvious implication of this aspect of complex system behavior is that there can be no guarantee of optimal efficiency as the system evolves: the state of the system is dependent on how it got there ("path dependent"), and it is always sensitive to future perturbations or changes in context.

4.2   THE AUTOMOTIVE TECHNOLOGY SYSTEM

A design engineer may view a product in isolation; in practice, the product's impact on society and its role in the economy cannot be so neatly circumscribed. For some products ("simple products"), their economic function and impacts depend primarily on the product's material composition (e.g., motor oil, windshield washer solvent). For others ("complex products" like the automobile itself), the product's function and impacts depend primarily on design, not materials. The approaches used to assess the environmental aspects of the design of one type of product will thus need to differ from those used for the other.

It is obvious that, like any fundamental technological system, the automotive technology system, encompassing both simple and complex products, is thus a complex system itself, and a rapidly evolving one at that. It is therefore appropriate to provide an overview of that system in terms of its most important levels, some of which are not usually considered when the environmental impacts of cars are decried. This approach is important in introducing analytical clarity and rigor to any discussion of a technology system. For example, despite the introduction of automotive technologies that have dramatically reduced emissions per unit, emissions as a whole have gone up because of responses in another part of the system (e.g., the cultural and psychological responses of consumers to advertising for four-wheel-drive sport utility vehicles and cheap gasoline, and living patterns that demand more automobiles per family unit and that they be driven long and often).

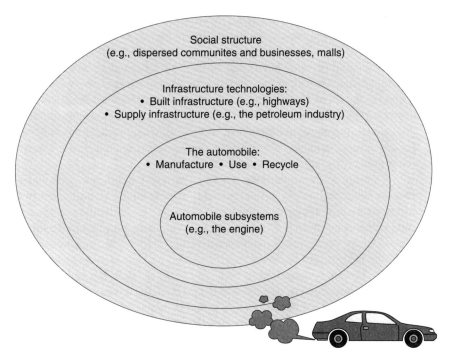

Figure 4.1   The automotive technology system: a schematic diagram.

Thus, further technological advance in engine design, for example, should not be expected to greatly improve air quality unless the customer use pattern subsystem is also addressed in some way.

Figure 4.1 is a simple schematic diagram of the automotive technology system. It includes the lowest, relatively technology-rich, levels of a car's mechanical subsystems and the manufacturing processes by which they are made. These subsystems and processes have been a predominant focus of environmental regulation, but, taken as a whole, they are probably not the major contributors to the automobile's environmental impact, except in a very local sense. For example, it is certainly reasonable to encourage the use of paints that do not contribute to local air pollution (e.g., paints that have low emissions of volatile organic compounds), but at the same time we should recognize that paint emissions are not a major environmental impact of the automobile.

The next level, automobile use, is more important. There are two major dimensions to this system level: technical and cultural. On the technical side, great progress in reducing environmental impacts has been achieved, and more is possible. Examples include sensor systems that monitor oil composition and properties and recommend oil changes only when needed, and sensor systems that report when air pressure in tires is low (low pressure results in higher gasoline consumption and greater tire wear). From the cultural standpoint, however, failure to address environmental impacts is virtually total. The mix of cars purchased in more developed countries—and soon to be available in rapidly developing countries such as China—is increasingly inefficient. In the United States, families routinely purchase vehicles with four-wheel-drive systems and high gasoline consumption. In addition, the gas-saving federal 55 mile per hour speed limit was repealed in 1995 on the grounds that it was an imposition on personal liberty and states' rights; depending on the state, legal speeds are now 55, 65, and 75, and, in one case, unlimited. In Germany, as was noted earlier, attempts to impose any speed limits on the autobahn routinely fail, even though the German environmental party (the Greens) is widely popular. Even a cursory evaluation of this simple systems model, therefore, indicates that, in this instance, much attention is being focused on the wrong subsystem, and illustrates the fundamental truth that a strictly technological solution is unlikely to fully mitigate a cultural problem.

Contrary to the usual understanding, the most significant environmental impacts of the automobile technology system probably arise from the highest levels of the system, the infrastructure technologies and the social structure. Consider the energy and environmental impacts that result from just two of the major infrastructures required by the use of automobiles. The construction and maintenance of the "built" infrastructure—the roads and highways, the bridges and tunnels, the garages and parking lots—involve huge environmental impacts. The energy required to build and maintain such infrastructure, the natural areas that must be perturbed or destroyed in the process, the amount of materials, from aggregate to fill to asphalt, demanded—all of this is required by the automobile culture, and attributable to it. Similarly, the primary customer for the entire petroleum sector—and, therefore, causative agent for much of its environmental impacts —is the automobile. Efforts are being made by a few leading infrastructure production firms such as Bechtel to reduce their environmental impacts, but these technological and management advances, desirable as they are, cannot in themselves compensate for the increased demand for infrastructure generated by the cultural patterns of automobile use.

The final and most fundamental effect of the automobile, however, may be in the geographical patterns of population distribution to which it has been a primary contributor. Particularly in lightly populated and highly developed countries such as Canada and Australia, the automobile has resulted in a diffuse pattern of residential and business development which is unsustainable without constant reliance on the automobile. Lack of sufficient population density along potential mass transit corridors makes public transportation uneconomical within many such areas, even where absolute population density would seem to augur otherwise (e.g., in densely populated suburban New Jersey). This transportation infrastructure pattern, once established, is highly resistant to change in the short term, if for no other reason than that residences and commercial buildings last for decades. Moreover, these trends do not seem to be improving. Data from the United States indicate not only that the number of commuters using public transportation in most urban centers is small, but that automobile commuting continues to increase. Thus, high demand for personal transportation (i.e., the automobile) is firmly embedded in the physical structure of the community.

Figure 4.2 illustrates some of the relationships among cultural and technological aspects of the automotive technology system, and relates these to the time frame in which they can be evolved. For example, the internal combustion engine is principally a technology system, and can be evolved relatively rapidly; the technology content of the infrastruc-

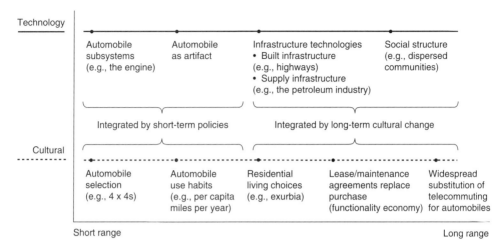

Figure 4.2   Relationships between cultural and technological aspects of the automotive technology system.

ture and the dispersed community structure, on the other hand, is relatively immune to evolution in the short term, principally because of the way it is embedded in the physical structure of the economy. Similarly, automobile selection can be evolved rather rapidly, even though it is a heavily cultural subsystem (after all, the shift to, then away from, energy efficient vehicles in developed countries has occurred within a twenty year time frame). On the other hand, evolving teleworking alternatives to commuting in an automobile have been actively encouraged by individuals and some governments for twenty years as well, with relatively little to show for it. Here, the technologies to support this transition have long been available; they simply aren't being used. In the short term, policy informed by industrial ecology must link culture and technology. In the longer term, it is likely that these two facets of the automotive technology system must be linked through cultural evolution.

This discussion illustrates the value of viewing the system within which one is operating, and provides one possible way of understanding the automotive technology system. It also illustrates the fallacy, usually based on an incomplete understanding of the system, of focusing solely on technological subsystems when, in fact, the cause of poor systems performance lies in the cultural or economic realm. This is not to say that technological evolution cannot serve an important role in mitigating environmental impacts even where cultural patterns—which usually require much more time to evolve—are involved. Indeed, DFE and other methodologies explored in this text attempt precisely that by taking a life cycle approach to automotive performance and design. Unquestionably, we are better off because automobiles are far more environmentally friendly today than they were in the past.

4.3 EVOLVING THE AUTOMOTIVE TECHNOLOGY SYSTEM

The recent evolution of the automobile in response to increasing concern about its environmental impact provides an interesting analogy to the path that must be followed by other technological systems as the global economy moves towards sustainability. The history of this evolution can be broken down into three periods. In the early 1960's, automobiles became ever more powerful in response to customer demand. As power increased rapidly, so did exhaust emissions.

In the late 1960's and early 1970's, two trends combined to render this technology state unstable. It became apparent that automobile emissions were, in fact, contributing substantially to unacceptable air quality in two primary ways: by the generation of photochemical smog and by the dispersion of lead into the environment (fuel additives containing lead not only lubricated the engine but allowed higher compression ratios than would otherwise have been possible). Moreover, in 1973 the so-called "gas crisis" struck, temporarily raising the cost of gasoline. This created strong, and to some extent conflicting, pressures on automobile manufacturers, who attempted to keep horsepower and performance levels high while simultaneously increasing miles per gallon and reducing emissions. In the short term this proved difficult, and horsepower per unit engine displacement suffered. By the late 1980's and early 1990's, however, sufficient time had elapsed to permit a re-engineering of the internal combustion automobile engine. The result was engines with high power and performance levels but much lower exhaust emissions.

There are other, equally significant, implications inherent in the evolution of the modern automobile engine system. Advances in engine performance have been achieved not by unexpected fundamental technological breakthroughs, but by the better application and integration of existing or incrementally improved technologies: use of composites and engineering plastics in the automobile body, which make the vehicle lighter; and more aerodynamic designs, which improve performance, for example. Technology diffusion, in other words, played a critical role. Information has been substituted for scarce physical resources and energy: examples include the sensor systems that now help to minimize emissions, and intelligent, computer-controlled engines that provide optimal performance, both being linked to existing road and driving conditions. Like the home or office, the engine has, in a fundamental sense, become a more efficient system by becoming a more complex system, with more feedback loops and information content built in, as Fig. 4.3 demonstrates.

The combination of these technologies has resulted in a truly better product: one that is more efficient in every way, yet at the same time one that provides greater quality-of-life (more durability, more safety, better handling, no dispersion of lead) to members of society.

This evolution provides an interesting, if not perfect, analogy to the stages a society must go through in its overall effort to integrate science, technology and environmental considerations in all economic activity. The first stage of this integration process was essentially a band-aid approach: environmental impacts were treated as completely ancillary to the primary economic activity: fast cars or industrial production. In the second stage, it began to be recognized that emissions must be controlled, but the underlying technological systems were not altered: this was the early 1970's car and the compliance stage of environ-

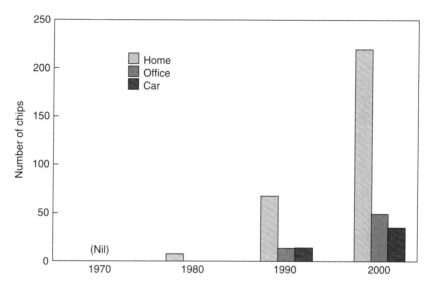

Figure 4.3   The rise of automated information management, as exemplified in the number of computer chips in office, car, and home from 1970 to 2000. (After a figure in Survey of computer industry, *The Economist*, Sept. 17–23, 1994, based on Motorola data.)

mental regulation. In either case, the blend of naive "end-of-pipe" control and pre-existing technology was an uncomfortable one, and produced less than optimal performance. The third and more desirable stage involved the re-engineering of the underlying technology so as to integrate economic, engineering, and environmental efficiency from the initial design stage. This approach has now become standard in the case of the car engine, but, in general, societies in few other situations have yet recognized the need for, and benefits of, the third stage. Consequently, most existing environmental policies and regulations remain analogous to the 1970's engine stage. This is understandable: that the task before us requires the complete re-engineering of the technological and societal systems characteristic of the Industrial Revolusion is daunting indeed.

The analogy is useful in suggesting future paths, however. It is optimistic in suggesting that, if the technology systems handed down as a result of the Industrial Revolution are, indeed, re-engineered, we can hope for substantial improvement not just in environmental performance but also in engineering and economic efficiency. The viability of at least the "weak sustainable development model"—better quality-of-life with less environmental degradation—is clearly supported by the automobile analogy. It suggests that substantial environmental improvement, with little or no cost penalty and possibly even increased quality-of-life, is possible with good engineering. At the same time, the evident limits to technological fixes are instructive: consumers and society as a whole must not be left with the impression that simply relying on technology will avoid the need for difficult and complex political decisions; technology has not, for example, made any significant impact on the endless traffic jams in all the world's major cities. Better technology can buy time, it cannot by itself buy sustainability.

It is precisely this non-technological dimension of the automobile that is so instructive to the industrial ecologist studying technology systems. The singular psychological and cultural appeal of the automobile—which is, unlike many artifacts, not limited to any region but is deep and global in scope—has no parallel in the modern economy. Any effort to understand the ecology of the automobile, or to regulate or modify its technological characteristics, fails unless the societal dimension of its existence is understood as a fundamental part of the automotive technology system.

SUGGESTED READING

Committee on Global Change, U.S. National Committee for the International Geosphere-Biosphere Program, *Toward an Understanding of Global Change*, National Academy Press, Washington, DC, 1988.

Costanza, R., L. Wainger, C. Folke and K. Maler, Modeling complex ecological economic systems, *Bioscience 43*, 545–555, 1993

Hofstadter, D. R., *Godel, Escher, Bach: An Eternal Golden Braid*, Vintage Books, New York, 1980.

EXERCISES

4.1 You are the head of the State Planning Commission of Utopia, a state which has previously been so poor that it now has no automobiles or infrastructure whatsoever. Due to a meteoric rise in the value of your primary product, utopia leaves, however, each of your citizens now has an income of $30,000 per year, and is naturally interested in achieving a developed country lifestyle. You have been charged with planning and implementing a sustainable automotive sector.

a. Define and justify several principles you would give your planning team to guide them as they create the highway network.

b. Identify and discuss the major costs and benefits arising from the impact of your new highway system on auxiliary systems (such as the economy, other technology systems and infrastructures, and living patterns and customs).

c. Assume that the sector will be completely Utopian; that is, every life cycle stage of the automobile, from manufacture to recycling, is performed only within Utopia. What is your plan?

d. Alternatively, assume that you must import the cars, and rely on the international energy markets and internationally available technology. What is your plan?

e. Assume two Utopias, one with the culture, population density, and population distribution of The Netherlands; the other with the characteristics of the United States. How do the plans you might develop under each set of conditions differ?

4.2 Many books have been written comparing biological systems with economic systems and the term "industrial ecology" reflects such an analogy between biological and industrial systems. How valid do you think such comparisons are? What are the benefits and potential pitfalls in relying on such comparisons?

4.3 It is often said that evolution of technological and cultural systems is a necessity if humans are to approach and maintain an environmentally and economically sustainable world. Develop a set of policies designed to encourage such evolution with respect to the automotive technology system. Integrate your policies into a time frame, from short term to long term, that leads to an environmentally preferable automotive sector in fifty years.

PART II: HISTORICAL PERSPECTIVES

CHAPTER

5

The Evolution of the Automobile

> I recall Roy Rogers and Trigger—an all-terrain vehicle with abundant onboard intelligence. Trigger always knew where he was, could find his way home if necessary, and understood moment-by-moment what his master needed; horse and cowboy functioned as one. But when the horse vanished from everyday life, leaving behind the horseless carriage, the onboard intelligence went too; there was a technological gap to be filled . . . Increasingly, now, electronics are doing the job. Soon, our automobiles will be at least as smart as Trigger, and the car-and-driver relationship will return to the cowpoke-and-horseflesh mode.
>
> W.J. Mitchell, *City of Bits:*
> *Space, Place, and the Infobahn*, MIT Press.

5.1 FROM HORSES TO AUTOMOBILES

It is far from straightforward to determine when the first automobile was made, because one must begin by defining the distinguishing characteristics of an automobile. Clearly it is a conveyance that carries one or more passengers, but so is a horse. Clearly it is a wheeled vehicle, but so are carriages and rail cars. Clearly it has its own means of propulsion, but so did steam-powered and electric vehicles. Perhaps the distinguishing feature, and the invention that made the modern automobile possible, is the internal combustion engine, applied to the propulsion of vehicles independently but more or less simultaneously around 1885 in Germany by Karl Benz and Gottlieb Daimler. Within a decade, a number of individuals and small companies in Europe and North America were manufacturing and selling automobiles. By consensus, the year 1996 is generally regarded as the centennial of the automobile.

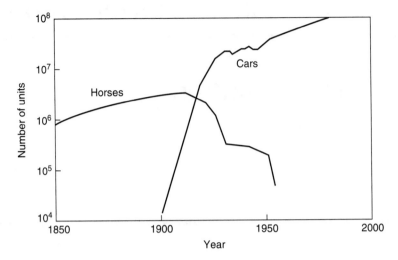

Figure 5.1   The evolution of transportation in the United States from 1850 to 1985: horses and automobiles. (Adapted from N. Nakićenović, Dynamics and replacement of U.S. transportation infrastructures, in *Cities and Their Vital Systems*, J.H. Ausubel and R. Herman, eds., Washington, D.C.: National Academy Press, 1988).

Automobiles were initially the toys of the wealthy, but two developments in the early 1900s changed this picture. One was the introduction of a variety of mass production methods: assembly lines, interchangeable parts, and conveyer belts. The second was the discovery of abundant oil in Texas. The result was dramatic decreases in the prices of both automobiles and the fuel on which they ran, and a rapid increase in the number of automobiles being made and purchased.

The initial growth in the number of motor vehicles was for the purpose of replacing horse-drawn vehicles. Figure 5.1 shows the relative numbers in the United States. By about 1915, automobiles had surpassed horses numerically. This latter growth phase was connected with the development of automotive infrastructure (roads, gas stations, etc.), and began an expansion and societal change that continues today. In fact, the world population of vehicles has grown to nearly 600 million (Fig. 5.2), about 25% of which are buses and trucks. As we will see in the next chapter, this implies an infrastructure of highways, fuel supplies, and sales and repair facilities that are in themselves very large in number.

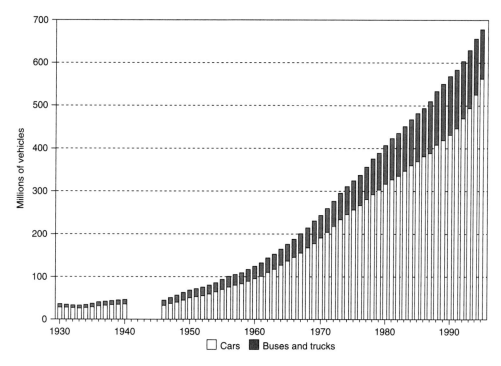

Figure 5.2   Size of the world automobile fleet (i.e., the number of vehicles registered). The data are from *World Motor Vehicle Data*, Detroit, MI: Motor Vehicle Manufacturers Association of the U.S., 1986, and various more recent references.

5.2 AUTOMOTIVE ASSEMBLY TECHNIQUES

The manner in which automobiles are assembled has evolved greatly over the years. Initially, every vehicle was made by hand, with components that would fit that vehicle only. Later, the assembly line was devised. It depended on the availability of parts that were interchangeable (i.e., that were so nearly identical that any one of them could be selected for use in any vehicle of the same model). Given bins of these parts, then, workers could install appropriate parts as vehicles moved to their station. In the initial embodiment of the assembly line technique, the vehicle chassis was wheeled from point to point. Henry Ford improved on that practice in about 1913, when he installed moving conveyer belts.

   Variations on the assembly line approach have been adopted for the manufacture of many products, automotive and non-automotive, and are still used today. A modern approach, using assembly teams, is something of a throwback to the early days of the automobile, since a small team is responsible for assembling most of the entire automobile. The result has been a significant improvement in worker satisfaction and in the quality of the vehicles produced.

5.3 THE DEVELOPMENT OF AUTOMOTIVE COMPONENTS AND SYSTEMS

This book does not pretend to be a treatise on automotive design. Nonetheless, the environmentally related attributes of the typical automobile require a working knowledge of the principal components of the product, some information concerning how they interact, and the ways in which they influence aspects of the environment. From those perspectives, we provide below brief descriptions of the parts and systems that constitute the modern automobile.

5.3.1 Chassis, Frames, and Engines

The first automobiles were built by carriage makers and had chassis that were virtual duplicates of vehicles designed to be pulled by horses. Made largely of steel, they were designed for rugged conditions and were heavy and lethargic in response to the wishes of the driver. The body frame might also have been of steel, but was more likely to be of wood. The strength and stability (today termed the "body integrity") of those frames was poor.

The modern vehicle has a chassis of high-strength steel designed for longevity and modest weight. Body members may be of either steel or aluminum. The intensive efforts to design a car with good performance characteristics, reliability, and low weight inevitably result in a close mingling of large numbers of diverse materials in the chassis and frame, a factor that complicates eventual disassembly.

The original automobile engine material was cast iron. Most of the related power system components—pistons, rods, drive shafts, axles—were and are of steel. In the 1960s engines began to be made from aluminum, a lighter-weight material, but one that is less rugged. The vast number of engine parts—covers, hoses, sensors, pulleys, valves, clamps—results in tens or hundreds of different materials in the makeup of a modern engine. Even ceramic engines have been proposed and fashioned, but none have gone into production. (However, many engines have ceramic components.)

5.3.2 Fuels

Gasoline and diesel fuel have been the predominant power sources for the motor vehicle ever since the late 19th century. In the 1930s, the development of engines with higher compression ratios than previously used necessitated the inclusion in fuel of antiknock components, of which the most effective was found to be tetraethyl lead (TEL). The TEL had the additional favorable attribute that it provided a degree of valve seat lubrication. In the 1970s, however, lead emissions were implicated in human toxicity problems. In addition, the catalytic converters developed about that time were poisoned by lead in the exhaust stream. As a result, lead in motor vehicle fuel has been virtually eliminated in many countries and decreased substantially in others. Many countries, however, mostly in early stages of industrial development, continue to use TEL.

Ethanol, mixed in various percentages with gasoline, is a fuel extender that has been used most extensively in Brazil, where its incorporation as a vehicle fuel began several decades ago. Prospects for more extensive use of ethanol are discussed in Chapter 9.

In the early 1960s, increased horsepower and performance were being obtained simply by increasing engine displacement (the summed volume of the cylinders) and compression ratio (the extent to which the air-gas mixture in each cylinder was being compressed before ignition). As a result, the measure of transport efficiency, distance traveled per unit of fuel consumed (i.e., km/l or mi/gal), dropped rapidly, and emissions of hydrocarbons, CO_2, and NO_x increased significantly. The 1960s automobile was efficient only if one assumed gasoline and emissions were essentially costless.

In the late 1960s and 1970s, the gasoline crisis and increasing air pollution stimulated the production of smaller engines with lower exhaust emissions, but the engines were far from optimum. By the late 1980's and early 1990's, however, sufficient time had elapsed to permit a redesign of the internal combustion automobile engine. The result was a triumph of engineering: a superior technology system in all ways compared to historical levels. Average engine size decreased by over 1500 cubic centimeters from 1974 to 1992 (Fig. 5.3), even while engine horsepower, after dipping during the late 1970s and early 1980s, returned to its old levels (Fig. 5.4). Thus, two trends were reflected in a dramatic increase in power per unit displacement over that time (Fig. 5.5) and an equally impressive improvement in vehicle performance, as measured in acceleration per unit time (Fig. 5.6). But the crowning achievement was to accomplish all this while both reducing emissions dramatically and increasing fuel economy substantially (Fig. 5.7). To top it all off, engines became more robust: in the 1960's, it was not unusual for engines to last for only 50,000–75,000 kilometers; now, well over 150,000 kilometers is routinely expected.

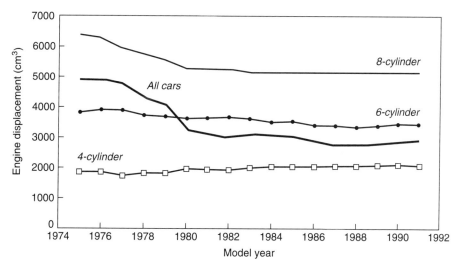

Figure 5.3   The average engine size for U.S. passenger cars, 1975–1991. (Reprinted with permission from *Automotive Fuel Economy*, Washington, D.C.: National Academy Press, 1992.)

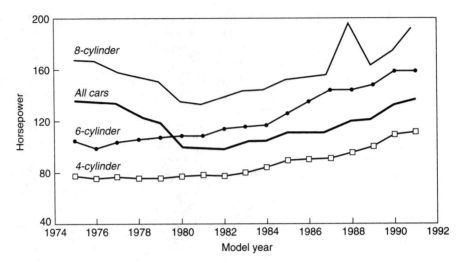

Figure 5.4   The average engine horsepower for U.S. passenger cars, 1975–1991. (Reprinted with permission from *Automotive Fuel Economy*, Washington, DC: National Academy Press, 1992.)

5.3.3   Tires

The wheels on the original automobiles were of iron strip connected through spokes to the axle; these were soon supplanted by solid rubber tires and, in the early 1900s, by pneumatic tires (tires filled with pressurized air). Until the 1950s, the air in tires was contained in inner tubes; inner tubes have since been eliminated in favor of tight seals between tire and wheel rim.

Natural rubber, strengthened by vulcanization (heating mixtures of sulfur and rubber to cross-link the rubber molecules), was the tire material of the first half of the 20th century. During the late 1940s and early 1950s satisfactory synthetic rubber began to become available, and today's tires are largely of synthetic rubber.

Modern tires contain more than rubber, of course. Fiber belting of tires to provide increased strength began in the 1970s and steel cord a few years later. Multimaterial tires are now standard; they provide much improved performance and have a number of optional end of life uses, but recovery of the original materials is generally infeasible.

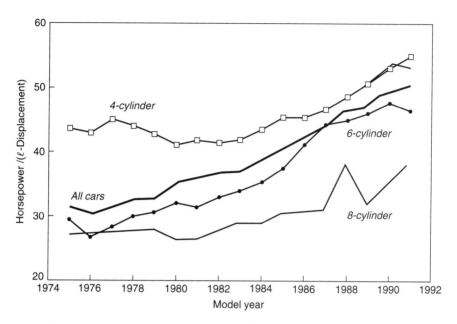

Figure 5.5   The average ratio of horsepower to engine displacement for U.S. passenger cars, 1975–1991. (Reprinted with permission from *Automotive Fuel Economy*, Washington, D.C.: National Academy Press, 1992.)

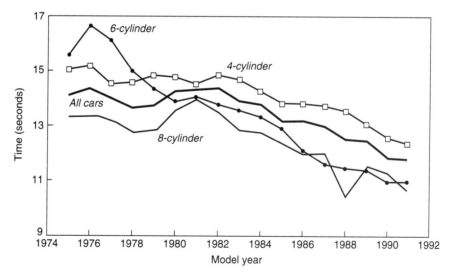

Figure 5.6   The average performance of U.S. passenger cars as measured by time to accelerate from 0 to 100 kilometers per hour, 1975–1991. (Reprinted with permission from *Automotive Fuel Economy*, Washington, D.C.: National Academy Press, 1992.)

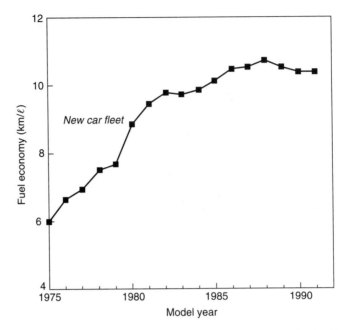

Figure 5.7   Trends in fuel economy for the U.S. new car fleet, 1975–1991. (Reprinted with permission from *Automotive Fuel Economy*, Washington, D.C.: National Academy Press, 1992.)

5.3.4   Brakes

Automobile brakes operate by the friction caused by rubbing two contact surfaces together. One surface is slowly abraded away as the braking action is accomplished. That surface has, throughout almost the entire life of the automobile, been asbestos, although concerns about the carcinogenesis of asbestos have recently led to the development of composite brake systems incorporating plastics and metal and glass fibers.

   Brakes were operated by simple foot pressure and cables until about 1930; that is, the brakes were wholly mechanical. Increased vehicle weight and the desire to make vehicle operation available to a wider spectrum of drivers resulted at that time in the development of hydraulic braking, in which the hydraulic system considerably amplifies the applied pressure. Automobile hydraulic fluids are largely glycols and glycol ethers.

5.3.5   Cooling Systems

Motor vehicle engines are cooled by a fluid circulating past the engine block and carrying heat away to a radiator (the latter usually made of copper alloy). An occasional vehicle is air-cooled (the 1960s-1980s Volkswagens being obvious examples). The original coolant was either ethanol or methanol, mixed with water as appropriate to provide protection against freezing. Alcohols tend to be excessively volatile, however, and they were gradually replaced in the late 1950s and early 1960s by ethylene glycol, again diluted as appropriate for the climate in which it was to function. Anticorrosive agents are customarily added to the glycol-water mixture, a complication that must be considered in recycling used antifreeze fluid.

To control the temperature and humidity within automobile passenger compartments, Oldsmobile introduced automotive air conditioners in the late 1950s. The working fluid was CFC-12 (CCl_2F_2), which performed well but was recognized by the late 1980s to be a factor in ozone depletion and global climate change. Beginning in 1993, new vehicles were supplied with air conditioners cooled by HFC-134a (CF_3CFH_2), a compound with substantially lower ozone depletion and global warming potentials.

5.3.6   Lubrication

The primary lubricant in motor vehicle engines from the turn of the 19th century until the present has been petroleum derivatives of suitable viscosity and composition. A number of additives—antioxidants, detergents, dispersants, corrosion inhibitors—enable oil to not only lubricate the engine but to clean it, reduce friction, and prevent rust. Motor oil is replaced when it becomes excessively dirty or when its constituents become less effective due to oxidation or other chemical breakdown.

Until the 1950s, a blow-by tube was standard design for the exhaust of volatile lubricant. Since that time, engine lubrication systems have been essentially self-contained. Once mostly expelled to the environment or discarded when changed, an increasing proportion of oil is today recovered, reprocessed, and reused.

5.3.7   Electrical Power

Electrical systems were originally installed on automobiles in 1912 to power electric starters. Once there, however, they soon made obsolete lights of kerosene and acetylene, as well as hand-operated bulb horns. Although alternatives have been tried, the standard energy storage technology for automobiles is the lead-acid battery. Together with the radiator, the alternator and cables inherent in the electrical system are the largest reservoirs of copper in the automobile.

Automotive electrical systems operate in difficult environments, and performance has been spotty over the years. A 1995 advance involving materials was the use of platinum tips on spark plugs. This material, together with improved control of the spark process, permits a single set of plugs to operate for as much as 6–8 years of normal driving.

5.3.8   Body Panels

Most body panels on the first vehicles and on those of today are of steel. A long-time exception was on station wagons and trucks, for which the load-carrying portion of the body was for decades made from wood, a practice that ceased in the 1950s. In the 1970s, however, the plastics industry developed patterned vinyl overlay material, and for about ten years most station wagons had steel sides covered with a wood-grain vinyl overlay. These overlays are now memories of the distant past.

The steel body panels on vehicles were traditionally protected by multiple paint coatings. In the early 1950s, however, corrosive road salt began to be used in abundance, and body panel corrosion increased markedly. As a result, steel panels are now commonly coated on one or both sides with zinc (this process is called "galvanizing").

Body panels made from polymeric materials made their appearance with the 1960s Corvettes, which used fiberglass. By 1983, glass-fiber reinforced thermosetting plastics and sheet molding compound were being used on a number of vehicles, especially passenger vans. Applications included hoods (bonnet lids), spoilers, trunk (boot) lids, front ends, spare wheel covers, air deflectors and spoilers. Polymer use for body panels has decreased slightly since about 1992, because of materials cost issues as well as the difficulty in recycling the composite materials.

Another body panel alternative to steel is aluminum, which is also being used in non-structural areas of automobiles. Lighter than steel and readily recyclable, aluminum has proved to be the material of choice for small panels and support members, but not for overall panel use, as high-speed production line welding of aluminum has proven difficult.

5.3.9   Bumpers and Body Trim

Bumpers have historically been made from heavy-gauge steel, and until the middle 1980s were chrome-plated. Other body trim, such as side-striping, was of chrome-plated steel as well. More recently, reinforced polymers backed by steel have been used for front ends and rear bumpers, and paint striping has replaced much of the metal body trim. As a result, there is now little chrome plating in the automobile industry.

5.3.10   Paint

Vehicle body panels have always been painted, both for aesthetic reasons and to minimize the corrosion of the panel itself. The paints were traditionally pigments contained in an organic solvent. At painting, the solvent was dispersed to the atmosphere and contributed to the development of photochemical smog. Beginning in the 1980s, efforts were made to develop satisfactory aqueous-based paints. As of the early 1990s, aqueous carriers had become widely used, and the mid-1990s saw the introduction of powder paints which do not require the use of any solvent at all.

5.3.11   Electronic Monitoring of Vehicle Performance

Electronic monitoring and control began to enter automobiles in the mid-1980s, and has now become an integral part of the automotive system. As with any electronics product, automotive electronics bring with them lead solder, copper-containing connectors, and mixed-material circuit boards. Recovery and recycling of the electronics components complicates the end-of-life process, but the materials can follow the burgeoning electronics recycling infrastructure.

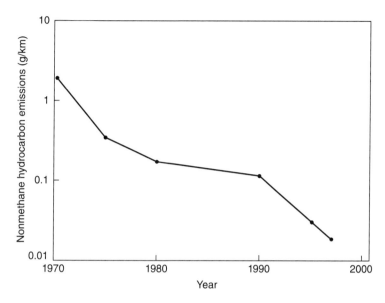

Figure 5.8   Non-methane hydrocarbon emissions from U.S. vehicles over the past three decades. The data are from the Low Emission Partnership, USCAR Consortium.

5.3.12   Emission Control Systems

Early automobiles emitted clouds of smoke from their pistons, as unburned fuel escaped freely to the atmosphere. In 1909, however, Frederick Lanchester of England developed oil-control piston rings, thereby sharply reducing "blowby" emissions of unburned fuel and lubricating oil. With evolutionary but not revolutionary changes, his concept survives to the present day.

The gradual realization that gaseous emissions from automobiles and other sources contributed to photochemical smog and other environmental hazards led in the early 1970s to the development of techniques to minimize those emissions. A major factor in emissions reduction has been the catalytic converter, in which catalysts of platinum, palladium, and rhodium convert unburned hydrocarbons and carbon monoxide to CO_2 and reduce oxides of nitrogen to N_2. As of 1989, about a third of all platinum-group metals consumption was for automotive uses. Electronic control of engine operation and electronic sensor monitoring of exhaust gases have also played major roles in emissions reductions, and have enhanced the ubiquity of electronics in the modern automobile.

The cycle of reductions in emissions of non-methane hydrocarbons (largely fuel molecules for which combustion was incomplete) is shown in Fig. 5.8. In 1970, emissions were about 2 g/km. In 1975, the introduction of exhaust catalysts reduced this rate by a factor of four. 1980 saw the first electronic sensors and exhaust system controls, and emissions dropped by an additional factor of two. The comprehensive use of precision technology in the 1990s—precise machining of combustion chambers, electronic control of fuel injection and ignition, and advanced exhaust gas sensors—has further reduced emissions, so that the vehicle of the late 1990s has hydrocarbon emission rates nearly a hundred times lower than those of 1970 vehicles.

Evaporative emissions are a related issue that has seen extensive improvement in recent years. Where vapor escape from fuel lines, carburetors, and gas tanks was once common (and was designed-in for pressure relief and to minimize the buildup of residues), evaporative emissions have been decreased more than 90% from 1970s-era levels by a variety of containment and vapor capture mechanisms.

5.3.13   Air Bags

In the late 1980s, automobile manufacturers began installing air bags on selected vehicles to protect passengers during collisions. By the mid-1990s, air bags were approaching the status of universal equipment. The principal material used to inflate air bags is sodium azide (NaN_3), which explosively decomposes upon impact, inflating the air bag virtually instantaneously. The material is harmless following bag deployment, but can damage shredding equipment if inadvertently left in vehicles that are being recycled.

5.4   USE OF MATERIALS

Throughout most of their history, automobiles have been largely constructed from a rather small palette of materials: steel, glass, rubber, and some sort of seat cushion material. In the late 1960s and early 1970s, other materials, the most significant of which was aluminum, began to come into use, and since about 1980 the materials in the automobile have become extremely diverse. Figure 5.9 follows the evolution of use by U.S. auto manufacturers of several materials over the 1980–1994 time period. Three patterns emerge from the figure. One, represented by steel and, to a lesser extent, by fluids, shows a gradual use decrease attributable to the Arab oil embargo of the mid-1980s and the public demand for smaller vehicles. With the lessening of concern for oil supplies, the first few years of the 1990s were marked by public demand for vehicles of increasing size and complexity, and total weight and hence steel and fluids use increased.

The second use pattern of Fig. 5.9 is that of glass and rubber, each of which shows a rather stable pattern of use, indicating little change in vehicle window area or tire size, on average.

The third pattern of Fig. 5.9, that for aluminum and plastics, shows a steady increase over the 15–year period. The pattern reflects two trends: the desire for lighter vehicles, and the progressive improvement in the properties, availability, and cost of these highly technological materials.

Another way to see how "engineered materials" have increased in use is to look at the degree to which the simpler cold-rolled and precoated steels have given way to better quality steels. As Table 5.1 shows, both high-strength steels and stainless steels are steadily increasing their fraction of overall automotive steel use.

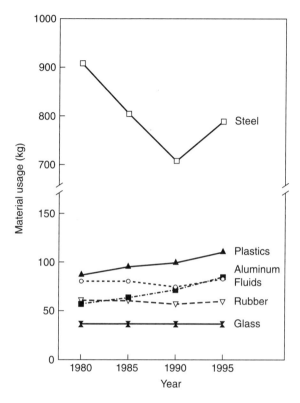

Figure 5.9   The use of different materials in the average automobile built in the U.S., 1980–1994. The data are from the American Automobile Manufacturers Association.

Table 5.1   Engineered Steels Use as Percentages of Overall Vehicle Weight of Steel

Material	1980	1985	1990	1994
High strength steel	8.8	12.4	14.9	15.1
Stainless steel	1.4	1.7	2.0	2.6

5.5 THE AUTOMOBILE AS A MODERN INDUSTRIAL PRODUCT

While automobiles are distinct products, many features of them are characteristic not only of other transportation products but also of industrial products of almost every kind and purpose. Among the features of automobiles we might cite in this connection are

- The product contains a wide variety of materials, many in close proximity and joined tightly.
- A high proportion of product parts and components is manufactured by outside suppliers and distributed as needed for final assembly. As a result, some portion of the total environmental impact of automobiles arises as a consequence of choices made concerning materials, industrial processes, and product delivery by these suppliers.
- The product is manufactured in facilities whose emissions to air, water, and land are closely monitored.
- The product combines electronics with hardware in an integrated "intelligent" system.
- The product is packaged and then shipped long distances to customers.
- An extensive recycling network exists for the product.

These characteristics will all be discussed later in this book from the standpoint of the automobile. In varying degrees, however, they represent as well the features of washing machines, lighting stanchions, soft-drink dispensers, elevators, aircraft, and most other products that have been or will be designed by industrial engineers. From that perspective, the automobile is not a technological system unique unto itself, but an archetype of modern technology.

SUGGESTED READING

Barker, R., and A. Harding, *Automobile Design: Twelve Great Designers and Their Work*, 2nd ed., Warrendale, PA: Society of Automotive Engineers, 411 pp., 1992.

Scharchberg, R.P., *Carriages Without Horses: J. Frank Duryea and the Birth of the American Automobile Industry*, Warrendale, PA: Society of Automotive Engineers, 243 pp., 1993.

K.T. Jackson, *Crabgrass Frontier*, New York: Oxford University Press, 1985.

EXERCISES

5.1 Braking, obviously a necessary function in an automobile, results in the loss of a lot of energy as heat. What ways can you think of to reduce this waste of energy? What other inefficiencies in energy use can you identify in modern automobiles?

5.2 Increasing demand for environmentally and economically efficient automobiles is generating cars that are more complex, contain more materials, and thus are harder to recycle than their predecessors. How would you balance better environmental performance against decreased recyclability? What criteria would you use to quantify such a comparison? If you were investing in research and development to reduce either or both of these impacts, what sort of research programs would you recommend?

5.3 Figure 5.2 gives the number of autos in the world over the period 1930–1990 and Fig. 5.9 gives materials usage per vehicle for the period 1980–1995. Compare the global amounts of the sum of all materials in the world vehicle fleet in 1980 and 1990, assuming that the data of Fig. 5.9 are representative of all vehicles. Comment on the results.

5.4   Figure 5.5 uses the mixed (English-metric) unit horsepower per liter of displacement. Devise a wholly metric unit for this vehicle parameter, and compute its value for 6–cylinder vehicles in 1975, 1983, and 1990.

5.5   Figure 5.1 gives the numbers of horses and cars in the U.S. from 1850 to the present. Assume the average vehicle horsepower to have been 50 in 1900, 75 in 1925, and 100 in 1950. Compute the total horsepower of horses and vehicles for those three years. Express the results in metric units. What percentage of the total in each of the years is attributable to horses?

CHAPTER 6

The Evolution of the Automotive Infrastructure

"Cities can be most simply described as being dense networks of communication, and the movement of persons is the critical process."
— Kevin Lynch, Massachusetts Institute of Technology

6.1 INFRASTRUCTURE AS A FOCUS

More than most other products of industry, the automobile requires an extensive and highly visible infrastructure for its use (Fig. 6.1). This infrastructure is of several types, the most obvious of which is the road and highway network, complete with bridges, tunnels, overpasses, entrance ramps, toll booths, and giant cloverleafs. A second major infrastructure component is the fuel extraction, transport, and delivery network, which involves oil wells in distant lands, supertanker transport of crude oil to refineries around the world, shipment of gasoline from refineries to petrol dealers, and sale to the customer. The third is parking lots and garages. A fourth is the network of automobile sales offices, designed so that potential customers can readily view and drive vehicles of interest. A fifth is the network of automobile repair shops, providing mechanical, electrical, and structural maintenance services to those requiring them.

This vast infrastructure requires us to think about the automobile as more than a self-contained industrial product. Its manufacture is really a part of a larger societal system, and in this sense it serves as a symbol for all industrial products. The designer of a washing machine, for example, assumes that there exists an infrastructure for providing energy, water, and detergent. The designer of a gas range assumes the existence of an infrastructure for providing natural gas. As these infrastructure components have been built up over the last half-century, it has been rare to engage in system-level thinking or planning, but one

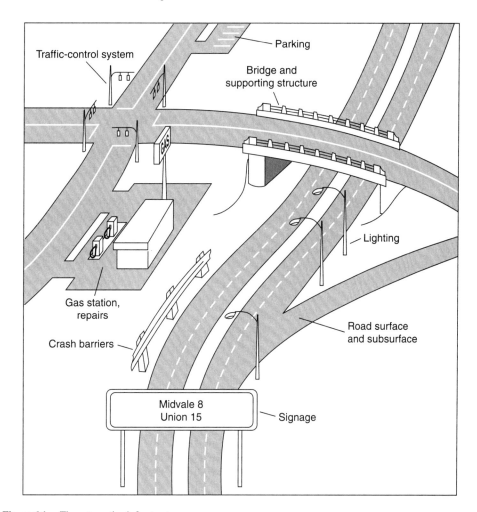

Figure 6.1   The automotive infrastructure.

could choose to do so, and might find that a system planned in such a way would be superior to one in which a product assumes a system, which is then built because the system assumes the increasing construction of similar products, which are then built because the infrastructure permits it, and on and on.

Those who envision how the world of tomorrow might look are sometimes termed "futurists". Futurists are able to design their pictures of societies of the future because they are not hampered in their thinking by what is in place today, as governments or product designers are. Inevitably, futurists design worlds in which systems thinking plays a central role. Waste material is eliminated in these worlds, because it is put to use in some way. Wasted time is minimized or eliminated as well, because the factors wasting time in today's world—traffic jams, queues at stores, etc.—are avoided in obvious or less obvious ways.

What is implied in these visions is the creation of new infrastructures to support the concepts of the new society. Before these can be created, however, we need to understand more about the current relationship between products and infrastructures. Infrastructure and how it has evolved in the case of the automobile is the central topic of this chapter.

6.2 DEVELOPMENT OF THE WORLD ROAD NETWORK

The world's first pathways were formed by animals traversing regular routes from one place to another. Many of these pathways are still in use today as roads, though their surfaces are different. Stone paving of city paths began as early as 4000 B.C. Over the next few millennia the roads extended out from the towns, brick was sometimes employed as a surface, and naturally-occurring bitumen (the viscous, heavy fraction of petroleum) began to be used as a mortar.

As with many aspects of civilization, road building achieved new heights under the Roman Empire. From 300 B.C. onward for six centuries, Roman roads ringed the Mediterranean Sea and crisscrossed all of Europe. In the process, Roman engineers developed concrete, multi-layer road substructure technology, and the routine shaping and use of paving stones.

Road construction and maintenance essentially disappeared for a thousand years with the decline of the Roman Empire. It was only when the growth of cities in the late Middle Ages required that substantial amounts of supplies be brought into cities, sometimes from great distances, that road building again became an important subject. In the 16th century, work in France, England, and elsewhere established that small stones of about 20 mm diameter were a satisfactory road surface if laid atop a base of stones 60–75 mm in size or larger. By the 17th century, such roads were common throughout Europe.

The 19th century resurrected an ancient but largely abandoned practice: the use of natural bitumen as a pavement binder. Originally employed in brick roads at various locations throughout the world, the bitumen was now used to bind sand and small stones into a mixture termed asphalt. Although early asphalt surfaces had many technical deficiencies, the surface was far superior to its predecessors, and from the early 1800s onward asphalt roads were in existence in the world's major cities.

The original automobiles were operated on the same dirt, stone, and asphalt roads used for horses, buggies, and the like, but the development of the rubber tire and of techniques for the construction of smooth road surfaces soon encouraged the system to evolve into an increasingly well-paved and interconnected structure. In the United States, for example, only about 5% of the roads were surfaced in any way in 1900 (Fig. 6.2). By 1990, two-thirds were paved, and nearly all the rest graveled or otherwise improved.

The construction of streets and highways involves the movement and use of more material than any other human activity except the construction of buildings. The material that is moved and used depends on the anticipated degree and intensity of use of the road. (The total distances of different categories of roads in the United States is shown in Fig. 6.3.) Roads receiving very modest use will normally have a gravel surface, the maintenance of which requires the periodic hauling of substantial amounts of stone and gravel. Roads

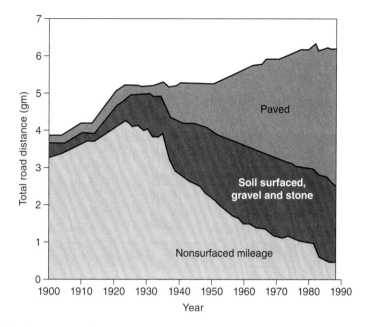

Figure 6.2   Total street and road mileage in the United States by type of surface, 1900–1989. (Adapted from *Transportation Statistics, Annual Report 1994*, Bureau of Transportation Statistics, U.S. Dept. of Transportation, Washington, D.C., 1994.)

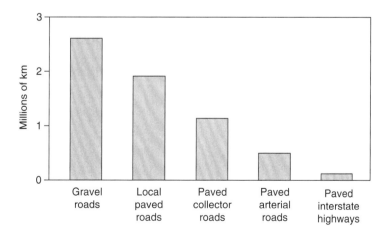

Figure 6.3   The 1990 U.S. road network, by type of road. The data are from *Highway Statistics 1990*, Federal Highway Administration, U.S. Dept. of Transportation, Washington, D.C., 1991.

Table 6.1 Average Materials Use per Million
US Dollars of Construction Cost

Material	Use
Aggregates	90 Tg
Cement	3.3 Gg
Bitumen	1.8 Gg
Reinforcing steel	0.4 Gg
Concrete, clay,	
PVC pipe	0.4 Gg
Lumber	17 Mg
Guard railing	4.9 Mg
Al culvert	260 kg
Fuel & lubicants	110 kl

Data from U.S. Federal Highway Administration,
Highway Statistics, 1994.

with more traffic, such as local and collector roads, are generally paved with asphalt. Arterial roads and expressways often have concrete surfaces.

Some feeling for the diversity and magnitude of the materials contained in a modern roadway is provided by Table 6.1. In addition to the anticipated items, large quantities of lumber are used as framing for concrete, culverts and pipes are incorporated routinely, and fuel and lubricant use for construction equipment is substantial.

The density of road networks has become a function both of affluence and of population density; it varies widely around the world, as shown by Table 6.2. Some countries,

Table 6.2 Roadway Distances in Selected Countries (km)

	Japan	Germany	Sweden	UK	Canada	Mexico	U.S.
Major roads (km)	5.8(4)	5.3(4)	1.6(4)	1.5(4)	1.5(5)	5.0(4)	3.3(5)
Secondary roads (km)	1.1(6)	5.8(5)	1.2(5)	3.5(5)	7.0(5)	1.9(5)	5.9(6)
All roads (km)	1.2(6)	6.3(5)	1.8(5)	3.7(5)	8.5(5)	2.4(5)	6.2(6)
All roads (km per 1K people)	9.0	7.8	15.6	6.4	30.4	2.7	24.2
All roads (km per km^2)	3.0	1.8	0.3	1.6	0.08	0.13	0.68

Data are for varying years from 1991–1993. 5.8(4) indicates 5.8×10^4.

Source: U.S. Federal Highway Administration, *Highway Statistics*, 1994.

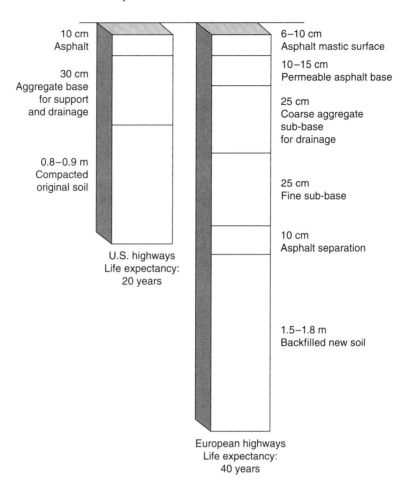

Figure 6.4 Typical construction designs and materials for major roadways in the United States and Europe. (U.S. Bureau of Mines).

Japan, Belgium, and The Netherlands, for example, have very high road densities, above 2 km of roads per square kilometer of area. Countries with low populations for the areas they encompass have low densities, below 0.1 km/km^2, especially if they are in the earlier stages of technological development. Affluent countries with moderate population densities generally have road densities somewhere between these extremes.

Approaches to road construction can be quite different. Figure 6.4 shows the contrast between typical roadways in the United States and Europe. The deeper road beds and intensive maintenance with surface asphalt allow European roads to last about twice as long as those in the United States. They are, of course, more costly and time-consuming to build. There is not necessarily a correct or incorrect way, but, as in most other things, Europeans take a longer-term view than do their North American colleagues.

6.3 BRIDGE DESIGN AND CONSTRUCTION

The first bridges were surely simple enhancements of fords by logs, or flat stones, which served as beams to bear the weight of animals or people crossing the waterway or chasm that was spanned. (This and other types of bridges are shown in Fig. 6.5.) The limit to the distance that could be spanned in this way was about 20 m, set by the size and load-bearing strength of the natural materials. The next development, which also was devised well before recorded history, consisted of using either natural stone islands or manufactured stone piers to permit a series of natural beams to span a larger distance. In principle, there is no limit to the length of such a bridge; in practice, the difficulty of constructing stable piers in deep chasms or in swiftly-flowing rivers limited the use of simple piers and natural spanning materials to relatively short, shallow streams and narrow ravines.

Another early bridge made of wholly natural materials was the suspension bridge, the rope being of vines or other natural fibers. Such bridges could span 50 m or more, and complex suspension bridges could include ropes for walkways and handrails, but they were unsuitable for animals or vehicles because of vibration and unstable decking.

Around 700–800 B.C., the keystone arch was developed as a method of bridge construction. Over the next several centuries, many of these bridges were constructed by the Romans to span distances of up to 40 m with a surface suitable for use by any type of road traffic. Keystone arch bridges required that the foundations be very secure and that extensive scaffolding be erected to support the stones of the arch as they were sequentially placed and fit during construction. From a planning and engineering standpoint, therefore, an arched stone bridge required substantial commitment and expertise.

Variations in pier and arch design occurred throughout the Dark and Middle Ages, with masonry arch bridge spans gradually increasing in length to 60–70 m. During the same period, however, the next advance in bridge design, the truss, was being developed. The truss, made of wood members, some in tension, some in compression, could span 60 m with an easily-worked material and without the necessity for massive stone foundations and piers.

Modern bridge building began with the wide availability of iron in the early 19th century. This led quickly to the construction of iron truss bridges. Such bridges could be twice the length of wooden spans, but their most common use was as strong, efficiently-constructed bridges spanning only 30–50 m. A more dramatic use of iron in bridge-building was the use of iron chains to replace ropes in suspension bridges, thus permitting still wider distances to be spanned. The Menai Straits Bridge in the U.K., completed in 1826, reached 174 m in length and had a deck that was fully road-worthy.

The final stage of bridge building to date was the development of the modern suspension bridge, which combined developments in iron and concrete. Concrete reinforced by iron bars was used to efficiently construct durable piers, stranded steel cables replaced the iron chains as the suspension members, and light, high-strength steel girders were employed for the towers and decking. Modern suspension bridges easily span 1000 m, and the Akashi Straits Bridge in Japan stretches 1780 m.

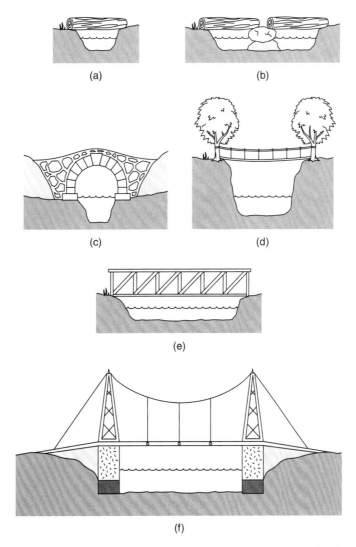

Figure 6.5   Bridging techniques, ancient to modern. (a) The natural beam bridge. (b) The two-span beam bridge, with a natural or constructed pier. (c) The simple suspension bridge. (d) The keystone arch bridge, with constructed foundations and crushed-rock fill. (e) The truss bridge. (f) The modern suspension bridge, with reinforced-concrete piers and steel towers and cables.

The modern expressway overpass is a bridge that generally spans 15–20 m. It is basically a simple steel beam construction on reinforced concrete piers, is relatively inexpensive, and can be erected rapidly. Only where spans are long, piers are high, or a multiplicity of gaps must be bridged, as in the intersection of two or more expressways, are the limits of the bridge designer's art employed, but the large number of short-span bridges on and over modern roads has had a significant influence on such environmental factors as new materials supply, disposal of obsolete bridge structures, and impact on local habitat.

6.4 THE AUTOMOTIVE SALES NETWORK

With some 12 million new vehicles and many more used vehicles being sold each year, automotive dealerships are among the more common of businesses. Each maintains a showroom or office of some sort, often quite large, and usually an adjacent area in which the stock of automobiles are stored and displayed. Precise numbers are difficult to come by, but it is estimated that there are some 100,000 new car dealers worldwide. Depending on the country, there are one or more dealers of used cars for every new car dealer, sometimes many more. Given an average ground area per dealership of about four hectares, the world's new and used car dealer showrooms and stock areas alone occupy at least eight hundred thousand hectares of land, or roughly the area of the Basin of Mexico, of which Mexico City, an urban area of more than 18 million people, is but a modest fraction.

6.5 THE AUTOMOTIVE FUEL DELIVERY NETWORK

Although a number of motive power sources for automobiles have been and continued to be used, by far the predominant power source is gasoline (petrol). In fact, some two-thirds by volume of all petroleum and petroleum derivatives used in the world are destroyed while providing motive power for automobiles.

In order to permit free use of the road infrastructure, gasoline must be widely and dependably available. Hence, an extensive infrastructure has developed for providing gasoline. The petroleum infrastructure worldwide includes drilling, pumping, loading, and transport facilities. Once crude oil reaches the country or region in which it is to be used, it is refined into jet fuel, gasoline, and a number of other products. The refineries are thus partly chargeable as automotive infrastructure. There are some 700 major refineries worldwide, producing about 17 million barrels of motor fuel annually (Table 6.3).

After the gasoline is produced at the refineries, it must be transported to the service stations at which it is sold to customers. This transport is normally accomplished by large tank trucks (which, of course, occupy roadways and consume fuel themselves.) A rough estimate of the number of petrol stations throughout the world is 400,000, or approximately four for every new car dealership. (The numbers are difficult to come by, because nearly all new car dealers offer maintenance services, as do many petrol stations, and some new car and used car dealers sell petrol. Rural farms and businesses often maintain their own petrol

Table 6.3 World Refinery Capacity and Output by Region

Region	Number of Refineries, 1994	Crude Oil Distillation Capacity	Motor Gasoline Output
North America	211	18440	8100
Latin America	76	5990	1050
Western Europe	124	14870	3380
FSU, Eastern Europe	83	12210	1430
Middle East	42	5050	650
Africa	46	2810	450
Far East, Oceania	125	13660	2370
World Total	707	73060	17420

The data are in units of thousands of barrels per day.

Data from U.S. Energy Information Agency, *International Energy Annual*, 1993.

supply. Depending on the statistical approach of the country concerned, a single establishment may be listed only once even if it performs three services, or it may be listed three times. All the numbers given here for the world's automotive-related businesses are therefore approximations.)

6.6 THE AUTOMOTIVE MAINTENANCE AND REPAIR NETWORK

As complex structures exposed almost constantly to high performance requirements and demanding environments, automobiles require substantial regular and on-demand maintenance. The network of shops set up to meet this need totals well over a million worldwide. In the U.S. alone, there were 334,000 auto repair shops in 1992. In addition, some 85,000 U.S. businesses provided auto-related services such as parts sales and car rentals.

In some regions of the world, automobiles are very long-lived and automotive maintenance is practiced even on vehicles two decades old or older. Those regions thus are heavy users of recycled and replacement parts, and are outstanding in inventing ways to keep very old vehicles on the road. This is a mixed blessing. Certainly, the rate of new vehicle production is much reduced. At the same time, the environmentally-related attributes now common in newer cars, such as low exhaust emissions rates and a need for less frequent oil changes, are absent. Without a careful and comprehensive analysis, it is difficult to tell whether the environment is better or worse off for older vehicles having their lives extended.

6.7   LAND DEVELOPMENT PATTERNS

Prior to the 20th century, urban areas were regarded as the height of culture and desirability. Their relative compactness enabled and encouraged such advances as water and sewer systems and public transportation. Commuter rail lines were beginning to make it possible to live outside the city and travel an extended distance to work, but such seemingly nonsensical behavior was uncommon. The automobile changed these patterns by eliminating the need for people to be concentrated near the rail infrastructure. The cost of this change was the creation of the highway infrastructure and, ultimately, what has become known as urban sprawl.

Until the late 1940s, automobiles were not commonly used for commuting to work. The rapid expansion of suburban housing in the United States, Canada, and some (but not all) other countries changed that pattern, however. Once housing was constructed on the road network rather than the rail network, automobiles became necessities for those living in suburbia, not only to get back and forth to work, but to shop, transport children, and visit friends. As shown in Fig. 6.6, about a third of all auto trips for work are today entirely in the suburbs, a pattern that results from the construction of offices and factories near roads but not near public transportation.

Given the need to use automobiles, and the rapid growth of suburban employment, highway construction has had to keep pace, and it has done so. Highways and other roads

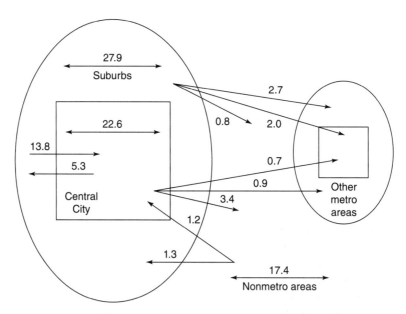

Figure 6.6   National flow patterns for people driving from home to work in the United States. The numbers represent percentages of all trips on the average workday. (Adapted from *Transportation Statistics, Annual Report 1994*, Bureau of Transportation Statistics, U.S. Dept. of Transportation, Washington, D.C., 1994.)

Table 6.4   The Use of Land in Morristown, NJ
by the Transportation Infrastructure

Land Use	Percent of Total Land
Roads, streets	8.2
Parking	1.8
Auto sales, new	0.2
Gas stations	0.1
Auto repair	0.1
Auto parts	0.06
Auto rental	0.04
Auto sales, used	0.02
Total	10.5

now define the boundaries and linkages of most urban and suburban areas; in central Los Angeles, some two-thirds of all the land is given over to the automotive infrastructure. A breakdown from Morristown, New Jersey, a suburban center of 17,000 inhabitants with moderate density residential areas, is shown in Table 6.4. The different infrastructure components are conservatively estimated to comprise 10.5 percent of the total land. This is a substantial fraction of the urban land to dedicate to a single use. It may indeed be what society wishes to do, but today's situation has certainly been a relatively unplanned happening, with the automobile encouraging the building of roads, thus the purchase and use of more automobiles, thus the need for parking and repair facilities, and so on. The inevitable result is that congestion and travel times increase (Fig. 6.7).

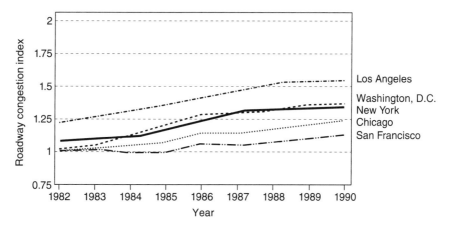

Figure 6.7   Roadway congestion indices from 1982 to 1990 in selected large metropolitan areas. (Adapted from *Transportation Statistics, Annual Report 1994*, Bureau of Transportation Statistics, U.S. Dept. of Transportation, Washington, D.C., 1994.)

A further aspect of urban sprawl and its enabling road network is the locations where the roads are built. Cities have historically been located near water, both because of abundant water supplies and because ships offered inexpensive and convenient transportation of people and goods. As extensive road structures were constructed to serve the people in those cities and those traveling to them, roads were inevitably built in environmentally sensitive areas such as swamps and wetlands. The consequences for natural vegetation and wildlife have often been severe, both because of loss of habitat and because of runoff of petroleum residues, road salt, and the like. The automobile and nature have not had a very comfortable joint existence.

This rather dismal picture is not restricted to the United States. Traffic jams, consumption of land, and disturbance of natural ecosystems have become a way of life in urban area after urban area, from Tokyo to Bangkok to Napoli to Mexico City. In many cases, the cause can be traced to the unrestricted use of automobiles in cities designed prior to the 20th century and in which no provisions for the combined occurrence in narrow streets of pedestrians, giant trucks, automobiles, motor scooters, and buses was ever considered. In others, even cities largely designed after the automobile was a part of everyday life, sheer numbers of vehicles complicate the most ambitious infrastructure program.

It has taken the better part of the 20th century, but especially the period 1950–1995, to produce the land use patterns encouraged by the wide availability of automobiles. It will take at least half a century to reverse those trends even should a plan be adopted and begun at once. This is not meant to suggest that the automobile is likely to be abandoned, but that the infrastructure should probably be modified to encourage the movement of people and things more efficiently than is now done. A mixture of mutually supporting infrastructure for public and private transportation seems advisable. It will take considerable commitment to bring it about.

6.8 INTERLOCKING INFRASTRUCTURES

The automotive infrastructure is a constituent of a larger system: the basic facilities, equipment, and installations necessary for the functioning of society. That overall system includes a number of other major components: the electrical power infrastructure, the water supply infrastructure, the sewer infrastructure, the communications infrastructure, and so on. The definition can also be extended to include natural systems on whose functioning society depends, such as the hydrologic system. Each of these components, in turn, can be subdivided; thus, we speak in the automotive infrastructure of the road and highway network, the fuel extraction, transport, and delivery network, the automotive maintenance network, and a number of others.

Infrastructure components are not independent. As a result, a major modification to or expansion of one infrastructure component often has positive or negative impacts on others. Some components are only weakly coupled: expanding the sewer network in a town, for example, has only a minor influence on the communications network. Other couplings are strong: major enlargements to parking lots often require storm sewer enhancements to handle the increased runoff volumes. Strongly coupled infrastructure components may therefore be termed "complementary components".

The impacts on complementary components may be interpreted narrowly or broadly. For example, one could evaluate a road built into the Amazon forest to support a new petroleum drilling operation from the standpoint of related physical infrastructure such as the need for water runoff piping. In fact, some petroleum companies have completely built and provisioned their Amazonian operations in such areas through the use of helicopters precisely to avoid such ancillary impacts. One could go further, and consider that once such a road is built it opens up the area to squatters and settlement, with the almost inevitable result that the biological communities in the area are severely disturbed. Similarly, if one designs a limited-access highway through a metropolitan area previously served by a typical urban thoroughfare, the result may be not just more rapid travel by the traveller, but urban decay as businesses in the now-bypassed center of town fail.

Some infrastructure interactions have extraordinary societal impacts: the first rail bridge over the Mississippi River, for example, forever made river shipping a secondary transportation option. (For this reason, the bridge was burned down twice, and barge operators came within a hair's breadth, amazing as it sounds now, of getting the United States Congress to pass a bill forbidding bridges over navigable rivers.) Conversely, building urban infrastructures that encourages redevelopment of brownfield sites and economic growth in inner cities may have significant positive externalities, which might well be recognized.

The automotive infrastructure turns out to have many complementary infrastructure components, as can be seen from Fig. 6.1. A list, not necessarily exhaustive, is given in Table 6.5. In several cases the potential impacts are anticipated to be large, either locally or regionally. While one cannot evaluate readily the entire implications of the modern technological society, it is nonetheless the case that, to the degree possible, assessing the environmental impacts of potentially important complementary infrastructure components should be a part of any assessment of the impacts of the primary infrastructure component under study.

Table 6.5   Complementary Infrastructure Components
for the Roadway Infrastructure

Component	Interaction	Potential Impact[*]
Electrical	Roadway lighting	Small
Sewer	Roadway runoff	Locally large
Telecommunications	Cellular calls	Moderate
Secondary roads	Roadway interconnections	Regionally large
Parking facilities	Vehicle parking	Regionally large
Petrol distribution	Vehicle refueling	Locally large
Bitumen distribution	Roadway maintenance	Regionally moderate
Natural hydrologic	Pulsed runoff	Locally large

[*]Potential impact of primary infrastructure modification or expansion of complementary infrastructure component.

SUGGESTED READING

Lay, M.G., *Ways of the World*, New Brunswick, NJ: Rutgers University Press, 401 pp., 1992.

Merritt, F.S., Ed., *Standard Handbook for Civil Engineers*, 3rd. ed., New York: McGraw-Hill, 23 chapters, index, 1983.

Stripple, H., *Livscykelanalys av väg, En Modellstudie för Inventering*, Report B 1210 (in Swedish), Göteborg: Institutet för Vatten-Och Luftvardsforskning, 1995.

Stammer, R.E., Jr., and F. Stodolsky, *Assessment of the Energy Impacts of Improving Highway Infrastructure Materials*, Report ANL/ESD/TM-115, Argonne, IL: Argonne National Laboratory, 1995.

Transportation Statistics, Annual Report 1994, Bureau of Transportation Statistics, U.S. Dept. of Transportation, Washington, D.C., 1994.

EXERCISES

6.1 Conduct a transportation land use survey of a 1 km^2 area of the business section of your community. Estimate the percentage of land devoted to the following: roads, parking, gas stations, auto repair facilities, auto sales offices, and other auto-related uses. Comment on the results.

6.2 Repeat Exercise 6.1 for two other modes of transportation: rail lines and bicycles.

6.3 Determine how much land must be used to support 1000 km of travel for one person by (a) foot, (b) bicycle, (c) automobile, (d) passenger train, and (e) commercial jetliner.

6.4 For the same transportation alternatives as in Exercise 6.3, determine how much petroleum is used per 1000 km per person, for both maintenance and power generation.

6.5 Cross-sections for typical roadways in the U.S. and Europe are shown in Fig. 6.4. What volume of aggregate would be required in each region to build a 4–lane undivided roadway (two lanes each direction) 10 km long, with lanes 5.5 m wide and shoulders (verges) 5 m wide? Repeat the calculation for asphalt, regarding the several types of asphalt as a single material.

6.6 Were methanol or ethanol to be used as the primary vehicle fuel instead of gasoline, a new and extensive fuel delivery network would need to be developed. The energy densities of these alternative fuels are less than that of gasoline: gasoline = 3.8 GJ/l, ethanol = 2.7 GJ/l, methanol = 2.0 GJ/l. The U.S. vehicle fleet travels some 4×10^{12} km annually, with average gasoline fuel efficiency of 10 km/l. What volumes of each fuel (in barrels) would have to be supplied in order to sustain that travel, assuming that the same total energy is needed no matter what fuel is used?

PART III: DESIGN FOR ENVIRONMENT

CHAPTER	**Choosing**
7	**Materials**

"It seems reasonable to view materials in process in industry, and the products of industry, more as transient embodiments of matter and energy in a flow of materials for human use than as 'wastes' with which we must deal."

— Robert Frosch, Harvard University

7.1 GUIDELINES FOR CHOOSING MATERIALS

The design of industrial products, whether automobiles, traffic signals, or highways, begins at least implicitly with the choice of the materials from which the products are made. Traditionally these choices have depended on the structural performance of potential materials, cost, esthetic appearance, ease of manufacturability, compatibility, and ruggedness. More recently, the impacts of materials choices on the local, regional, and global environment have been factors in materials choice. From that standpoint, several guidelines constitute a generalized approach to materials selection:

1. Choose abundant, nontoxic, nonregulated materials if possible. If toxic materials are required for a manufacturing process, try to generate them on site rather than to have them formulated elsewhere and shipped.

2. Choose materials familiar to nature (e.g., cellulose) rather than those with which nature has no experience (e.g., chlorinated aromatics).

3. Choose materials for which recycling at end of life is a reasonable option and for which a recycling infrastructure exists.

4. Design for minimum use of materials in products, processes, and in product use.

5. Acquire materials from recycling streams rather than from virgin material extraction.

These guidelines will be helpful as we assess the materials generally used in the manufacture of vehicles and in the creation and maintenance of the vehicular infrastructure.

7.2 THE LIFE CYCLE OF AN AUTOMOBILE

An approach that we have implied concerning one's concept of a particular product or system, but have not heretofore specifically stated, is that the entire life cycle should be considered. That is, one should consider the extraction from their reservoirs of the materials that are used, what happens to them (and the environment) during product manufacture, how the use of the product or system affects the world within which the use occurs, and, finally, what happens to the product or system and its materials once it is obsolete or the consumer disposes of it.

The life cycle of an automobile can be put into perspective with the help of Fig. 7.1. The cycle begins with the extraction of raw materials either from virgin sources or from a recycling stream: petroleum, coal, metal ores, rubber, and so on. The raw materials are then processed by a variety of industrial techniques to produce finished materials suitable for use in manufacture. During manufacture, another series of processes bend and shape the metal, mold the plastic, and perform the other operations needed to create components and assemblies from the finished materials. Product completion is followed by packaging, delivery, and customer use, generally for extended periods of time. A variety of maintenance activities require additional material streams during consumer use: lubricants, coolants, tires, etc. When the vehicle is finally retired, an extensive recycling operation restores a high percentage of the materials back into the industrial materials supply system. The entire cycle for a typical vehicle takes 10–14 years.

Similarly, we can construct a diagram (Fig. 7.2) for the life cycle of the automotive infrastructure system. The form of Figs. 7.1 and 7.2 is obviously similar, though there are differences in processes and products. Another difference is the time scale involved for the infrastructure life cycle. It is obviously much longer than for an automobile; bridges have lifetimes of perhaps half a century or more, and many road surfaces are periodically repaired and used essentially forever.

Thus, all products and systems can be perceived and evaluated from the viewpoint of their life cycle. There are several crucial insights that result from such an approach. One is that the designer should view the product or system from the standpoint not of "out of sight, out of mind," but over a period that, including the customer use and end of life stages, may encompass decades. Another is that one should consider the life-cycle impacts of the choices of materials, since their extraction and processing stages can have very large environmental impacts. A third is that the end of life stage can be very much enabled or disabled by the decisions made by the design team concerning which materials are selected or which assembly techniques are used.

The perspective of the life cycle is the central tenet of applied industrial ecology, and is the approach that should be taken to every activity of the industrialized society. In this book, we apply the tool to the automobile and the infrastructure that supports the automobile, but we emphasize that the tool of life cycle assessment is as universal as a screwdriver

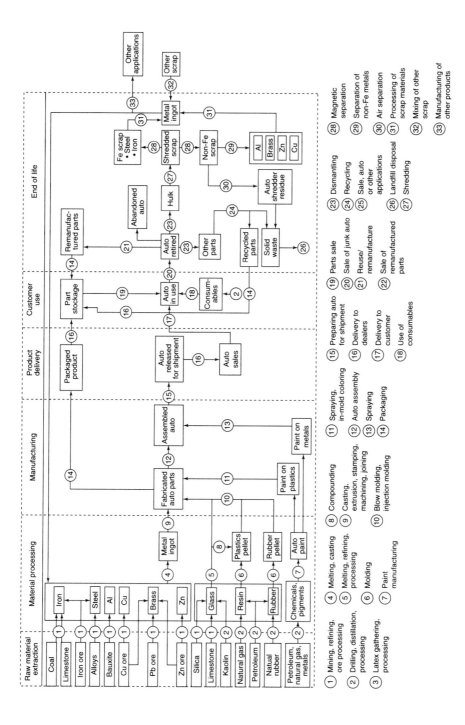

Figure 7.1 The life cycle of the automobile, and the processes that occur during that cycle. The processes listed at the bottom of the chart are keyed by number to the life cycle steps shown in the flow diagram above. (Adapted from a diagram devised by G. Keoleian, Univ. of Michigan.)

79

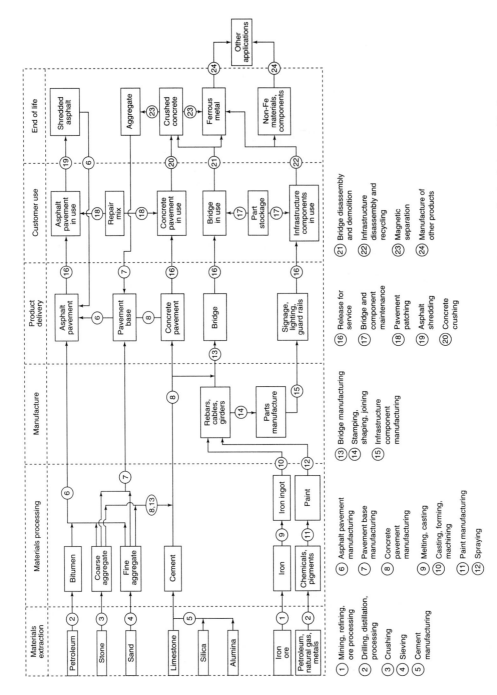

Figure 7.2 The life cycle of the automotive infrastructure, and the processes that occur during that cycle. The processes listed at the bottom of the chart are keyed by number to the life cycle steps shown in the flow diagram above.

80

or a hammer, and at least as valuable. Indeed, we anticipate that a toolkit of life-cycle assessment methods applicable to different tasks will be evolved over time, though few of those tools have thus far been developed.

7.3 MATERIALS IN A TYPICAL AUTOMOBILE

The modern automobile is a complex product containing a substantial diversity of materials. Table 7.1 gives average contents of materials for generic automobiles of the 1950s and 1990s. Substantial change in materials use has occurred during the 40 year interval, with modern vehicles using significant amounts of plastics and aluminum in addition to all of the materials used in 1950 (i.e., materials diversity has greatly increased). In overview, the 1950s vehicle was substantially heavier, was less fuel efficient, was prone to greater dissipation of working fluids and exhaust gas pollutants, and had components, such as tires, that were less durable. From a systems perspective, however, there were far fewer automobiles, and each was driven far fewer kilometers per year. Moreover, the increasing diversity of materials in a modern vehicle, while it enables far greater efficiency, may complicate recycling activities.

Let us see to what extent the materials choice guidelines given above are followed in the design of a typical automobile. The first guideline is for abundant materials. On that score, the automobile fares reasonably well, since iron (the major constituent of steel), aluminum, and silicon (the major constituent of glass) are among Earth's most abundant elements. Lead, zinc, copper, and tin used in smaller amounts, are, in contrast, among the elements whose supply is potentially limited. Plastics, synthetic rubber, antifreeze, and oil are all products related to petroleum, itself increasingly mentioned as the single most limited industrial resource. Given abundant energy, however, all could probably be produced from renewable biomass.

Table 7.1   Materials Use (kg) in Generic Automobiles (estimates from Ward's Automobile Yearbook)

Characteristic	ca. 1950s Automobile	ca. 1990s Automobile
Plastics	0	101
Aluminum	0	68
Copper	25	22
Lead	23	15
Zinc	25	10
Iron	220	207
Steels	1290	793
Glass	54	38
Rubber	85	61
Fluids	96	81
Other	83	38
Total Weight:	1901	1434

Several of the materials are moderately to highly toxic. Lead (in batteries and, to a small degree, in electronic circuitry) is perhaps of most concern. Mobilizable copper is also an environmental toxicant, but most of the automotive copper is in parts and components that commonly enter the remanufacturing stream. Antifreeze is widely known as hazardous, and some of the constituents of lubricating oil are harmful as well. Toxicity concerns are basically absent for steel, aluminum, glass, and plastics, except to the extent that they may contain toxic contaminants (e.g., small amounts of cadmium in tires).

The materials in modern automobiles vary greatly in their ability to be recycled. Steel and aluminum are good choices in this regard, though they typically suffer some degradation during recycling. Lead (in batteries) and copper (in radiators and in motor armatures) also have good recycling records. The situation is much poorer for plastics. Composites and thermosets are difficult or impossible to recycle, although they can safely and cleanly be burned to recover a portion of their chemical bond energy. Thermoplastics and other materials now becoming available have better records in this regard, but much needs to be done to enhance this effort. Another difficulty occurs when incompatible plastics are used together in component designs that make separation difficult or impossible. The intercompatibility of the most common plastics is shown in Fig. 7.3, and designers should use the information in the figure as a guide when incorporating plastics in their products. Used tires have been experimentally processed back into petroleum, and the rubber in tires is suitable for incineration. Glass and other miscellaneous materials are almost always discarded.

The philosophy of recommending the use of natural rather than synthetic materials is that any residues that escape will encounter ecosystems that have learned (by evolution over the eons) to deal successfully with them. In contrast, no biological defenses may have been developed that are equipped to deal with newly synthesized materials. Carefully-selected natural materials are often good humidity regulators, insulate well against heat and noise, are strong for their weight, and are generally inexpensive. As a result, natural materials are increasingly seen as suitable choices for the automotive design engineer.

Dematerialization, the use of lesser amounts of materials to accomplish the same functions previously accomplished by larger amounts of materials, has been an active part of automotive design for two decades or so, resulting in the decrease in materials use shown in Table 7.1. As we will see, it is not clear that designers can continue this record of achievement, but accomplishments to date have been reasonably impressive. It is, of course, necessary to evaluate dematerialization over the life cycle of the material to ensure that impacts have not simply been shifted to another life-cycle stage (e.g., aluminum processing is much more energy-intensive than steel processing).

Finally, how much of the material used in automobile manufacture is acquired from recycling streams? The surprising answer is only about 5%. Even though steel is eminently recyclable, recycled steel is currently too impure for the high-strength requirements of many automotive components, and is instead incorporated into lower-grade items such as castings or non-automotive products. Additional efforts on purification of recycled steel are obviously desirable. Aluminum, in contrast, tends to be horizontally recyclable; that is, it can be recycled back into the same use, rather than into another use which tolerates a more degraded material ("vertical recycling"). In fact, recycled aluminum constitutes some 30% of the incoming automotive aluminum flow stream. (Horizontal recycling is when a partic-

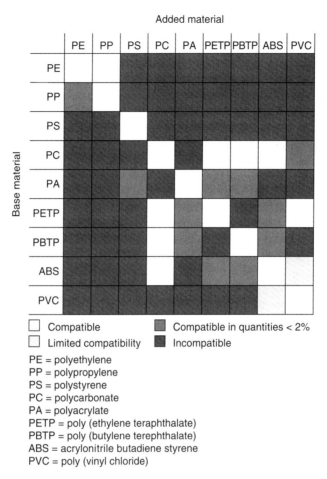

Figure 7.3   A matrix indicating the compatibility for recycling of different combinations of common plastics. The abbreviations are: PE, polyethylene; PP, polypropylene; PS, polystyrene; PC, polycarbonate; PA, polyacrylate; PETP, poly (ethylene teraphthalate); PBTP, poly (butylene terephthalate); ABS, acrylonitrile butadiene styrene; PVC, poly (vinyl chloride). (Adapted from Rogers, K., *The Motor Industry and the Environment*, London: The Economist Intelligence Unit, 1994.)

ular mass of material is remanufactured into the same product, as opposed to vertical recycling, in which a product from a particular mass of material is recycled into a different type of product.) The lead in lead-acid batteries and copper in radiators and motors are both part of very efficient recycling streams: some 90% of them are retrieved and returned to use. Occasional additional examples occur here and there, as in Ford's use of carpet from its offices to make mineral-reinforced nylon engineering polymers for its vehicles.

7.4 MATERIALS FLOW FOR A TYPICAL AUTOMOBILE

What happens to the materials in a typical automobile? That question was addressed in detail by Douglas Ginley of the U.S. Bureau of Mines, who studied the major materials in a typical 1990 automobile. His diagram for steel is shown in Fig. 7.4. A few data are of particular interest. For example, some 10% of the virgin material supplied for extraction is lost in processing. Nonetheless, the steel used in the vehicle is 95% virgin. Losses as a percentage of material decrease at each succeeding manufacturing stage. The reuse amount at the product use stage relates to those portions of the vehicle that are salvaged relatively intact: transmissions, alternators, and the like. The recycling stream from that stage represents separable components injected directly into a recyling stream: batteries, bumpers, and the like. The product disposal stage is, in contrast, the shredding stage, where the mostly recyclable metals are separated from the mostly unrecyclable other materials (glass and rubber are examples).

Figure 7.5 shows the flow diagram for automotive aluminum. The pattern has much in common with that for steel, for both metals are eminently recyclable. As with steel, about 5% of the aluminum entering the product disposal stage is lost to landfilling.

The flows for copper are given in Fig. 7.6. Among the more interesting features of this diagram is the high rate of component recycling following the product use stage (mostly motor and starter armatures and radiators) and the loss of a significant quantity of material during processing.

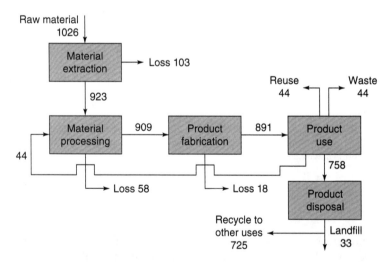

Figure 7.4   The flow of steel (kg) in a typical 1990 automobile. Reuse refers to actions that do not involve a materials processing step, such as the refurbishment of a transmission casing for use on another vehicle. (Adapted with permission from Ginley, D.M., Material flows in the transport industry, an example of industrial metabolism, *Resources Policy, 20,* 169–181, 1994.)

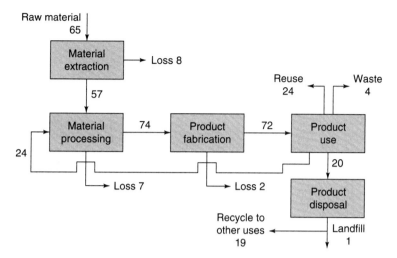

Figure 7.5   The flow of aluminum (kg) in a typical 1990 automobile. (Adapted with permission from Ginley, D.M., Material flows in the transport industry, an example of industrial metabolism, *Resources Policy, 20,* 169–181, 1994.)

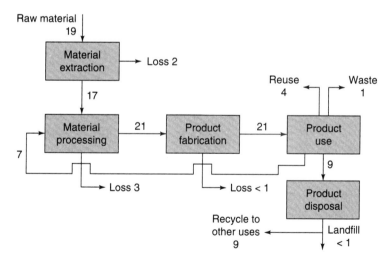

Figure 7.6   The flow of copper (kg) in a typical 1990 automobile. (Adapted with permission from Ginley, D.M., Material flows in the transport industry, an example of industrial metabolism, *Resources Policy, 20,* 169–181, 1994.)

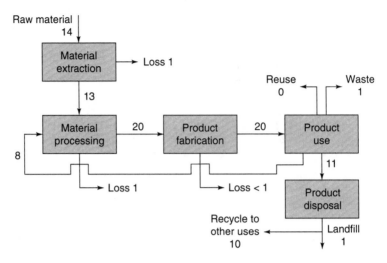

Figure 7.7 The flow of lead and zinc (kg) in a typical 1990 automobile. (Adapted with permission from Ginley, D.M., Material flows in the transport industry, an example of industrial metabolism, *Resources Policy, 20,* 169–181, 1994.)

Flows of lead and zinc are shown in Fig. 7.7. The lead, largely used in batteries, constitutes the bulk of the product use recyling flow. In contrast, the zinc is almost entirely used as an anti-corrosion plating for the steel body panels. It is not lost to landfills upon disposal; much is recovered as zinc dust in the foundries, and some reappears as an impurity in the reprocessed steel. Hence, the latter contributes to a degraded product and cannot itself be reused as zinc.

Plastics, an increasing proportion of modern automobiles, have typical flows as shown in Fig. 7.8. At this stage of the technology, they differ markedly from the metals in that they are almost totally lost to landfilling rather than being recycled. This fate occurs because (1) many automotive plastics are thermosets, a class of materials for which routine recycling technologies do not yet exist, (2) automobiles have very large plastics diversity, and (3) designers have not generally designed vehicles with plastics recovery and recycling in mind. Substantial efforts to reduce plastics diversity and to use more recyclable plastics are underway in most automobile manufacturing corporations.

Figure 7.9 shows the flows of mixed materials (glass, rubber, paper, etc.). Some of this material, especially tires, is salvaged and recycled after product use. Most, however, cannot now be retrieved and are landfilled. As with plastics, design for recycling can achieve major gains here. An example of what is possible is the increasing use of natural organic materials by Daimler-Benz, who employ coconut fiber for headrests and flax and sisal fibers for seat cushions and rear shelves on Mercedes sedans. These materials avoid the direct use of petroleum or other industrial resources, and have no toxic characteristics. Equally important, when these materials decompose or are incinerated, the carbon dioxide they release will be that which they absorbed from the atmosphere as they grew, so no additional greenhouse gas warming is implied by their use.

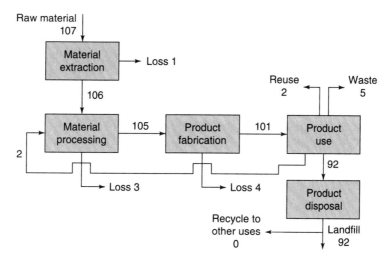

Figure 7.8   The flow of plastics (kg) in a typical 1990 automobile. (Adapted with permission from Ginley, D.M., Material flows in the transport industry, an example of industrial metabolism, *Resources Policy, 20,* 169–181, 1994.)

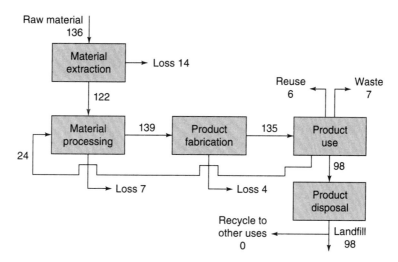

Figure 7.9   The flow of miscellaneous materials (rubber, glass, paper, etc.) (kg) in a typical 1990 automobile. (Adapted with permission from Ginley, D.M., Material flows in the transport industry, an example of industrial metabolism, *Resources Policy, 20,* 169–181, 1994.)

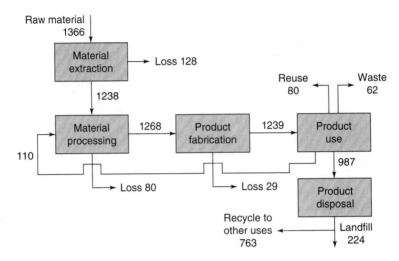

Figure 7.10   The total flow of materials (kg) in a typical 1990 automobile. (Adapted with permission from Ginley, D.M., Material flows in the transport industry, an example of industrial metabolism, *Resources Policy, 20,* 169–181, 1994.)

    The composite flow diagram comprises Fig. 7.10. Recycled flows for materials used in automobile manufacture are seen to be only about 10% of input materials overall (110 kg out of 1348). Losses during fabrication are small. Nearly 80% of the vehicle is recycled in some way, but much of the material can only be used in degraded form. About 15% of the raw material input is eventually landfilled.

    It is also of interest to examine the use of materials in automobiles as fractions of the use of those materials for all purposes. Table 7.2 provides this information for the United

Table 7.2   Materials Use by the United States Automotive Industry, 1992

Material	Automobile Use	Total US Use	Automobile Percentage
Aluminum	1.3 Tg	6.8 Tg	18.9
Copper	0.3 Tg	3.0 Tg	10.0
Cotton	4.8 Gg	2.2 Tg	0.2
Iron	16.8 Tg	48.6 Tg	34.5
Lead	0.86 Tg	1.24 Tg	69.5
Plastic	1.0 Tg	30.7 Tg	3.2
Platinum	26.4 Mg	63.7 Mg	41.4
Rubber	1.80 Tg	2.86 Tg	62.9
Steel	11.3 Tg	83.6 Tg	13.5
Zinc	0.3 Tg	1.2 Tg	23.0

States. In several cases, automotive use is the dominant flow of the material: lead (for batteries), natural and synthetic rubber (for tires), platinum (for exhaust catalysts). In contrast, even though plastics use in automobiles has been growing rapidly, it still is a tiny fraction of all plastics uses. For many other materials, including aluminum, steel, iron, copper, and zinc, automotive use is 10–35%; very important for the resource budget sheet, but not dominant.

In summary, the automobile is composed of a great diversity of materials, some much more amenable to reuse than others. Materials recovery and reprocessing is extensive but tends to be vertical (the material goes to a less sophisticated use) rather than horizontal (the material is returned to an equivalent use). Automotive designers need to devote increased attention to design for disassembly and to using more easily recyclable plastics. Recyclers need to expend more effort on the recycling of "process fluff", the residue of vehicle shredding that includes glass, trace amounts of metals and plastics, and other mixed materials. In this connection, materials scientists need to plan the properties of shredder residue so as to make it economically attractive for recyclers to recover and reuse it. They also need to develop techniques to produce high-grade recycled material that can be reused in new automobiles rather than relegated to lower-grade uses. Until high-grade recycled material can be produced, designers will be forced to go to virgin streams for satisfactory materials, a practice that continues the cycle of discarding of materials rather than reusing them over and over.

7.5 MATERIALS IN THE TRANSPORTATION INFRASTRUCTURE

Figure 6.1 illustrated the many parts of the automotive infrastructure. In considering the use of materials that the infrastructure's existence implies, let us look at the road itself, the infrastructure component that uses by far the largest proportion of materials. A schematic diagram of the typical "built" road is shown in Fig. 7.11.

There are three components to the road itself: the subbase, the base, and the surface layer. The subbase is the native soil or rock on which the road is constructed. It is sometimes combined with recycled material such as fly ash or crushed concrete. If the native material is not suitable, as might occur, for example, if an habitually moist area is being crossed, subbase material may need to be imported, an operation with high transport costs and environmental impacts. The subbase material is prepared as required by controlled blasting of bedrock, spread by bulldozers and rollers, and compacted into uniform layers of high density by rolling and tamping. Often the subbase material is stabilized by the addition of cement, bitumen, or lime.

The intermediate or base layer consists of crushed stone and gravel, normally spread to a total depth of 0.3–0.5 m or more, and providing both support and drainage. Depending on the composition of the subbase and the expected use of the road, a second base layer may be laid. Preparation and transportation of base layer material is common, especially if the road traverses sandy or loamy areas where rock is not readily available.

The most common type of road surface is asphalt, for which the key ingredient is bitumen, a semi-solid mixture of heavy hydrocarbons. Bitumens are occasionally found in

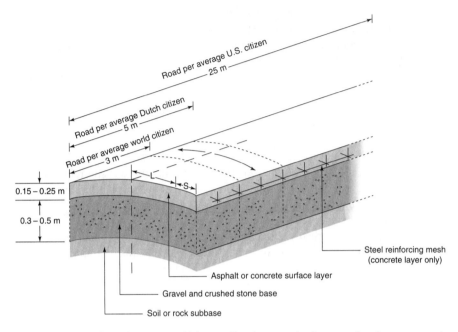

Figure 7.11   The structure of a modern concrete highway. Also shown are the distances of roads per average citizen of the United States, The Netherlands, and the world. Data for the latter calculations are from *World Road Statistics 1985–1989*, International Road Federation, Geneva, 1990.

nature, as in the LaBrea tar pits in Los Angeles, but are most commonly produced as residue from petroleum distillation. When bitumen is heated and mixed with small stones (termed "aggregate") and dust, the result is asphalt, a reasonably inexpensive, rugged, dark-colored substance that sees wide use in road surfacing. A typical mix of asphalt materials is given in Table 7.3; the result is a substance with bulk density of about 1.2 g/cm^3. A typical surface layer of asphalt is about 8–15 cm in thickness. Some asphalts are made with a few percent of ground tire rubber, which costs slightly more but produces a roadway with improved characteristics and less environmental impact.

The second common pavement material is concrete, which has Portland cement as the solidifying material. The cement is a mixture of lime, alumina, silica, and iron oxide. Con-

Table 7.3   A Typical Mix of Materials in Asphalt Pavement

Material	Weight percent
Coarse aggregate (2–3 cm)	50
Fine aggregate (1–2 cm)	38
Filler (mineral dust)	7
Bitumen	5

crete is manufactured at the construction site by combining cement, aggregate, sand, and water. While the latter three substances may have relatively small embedded environmental impacts, the manufacture of cement is very energy intensive. In fact, cement production contributes about 3% of the global annual production of carbon dioxide. The standard practice is to reinforce the concrete with iron bars or mesh. In this application, the wire support is placed atop the base layer and the concrete poured onto it, the grids serving as an interior support structure. Periodically, a polymeric expansion strip is inserted to separate pavement segments.

Estimates of the materials now in place in the transportation infrastructure are crude, but the magnitudes are so enormous that the numerical accuracy is secondary to the usefulness of order of magnitude assessments. These are provided in Table 7.4. An interesting perspective is that a few meters of highway can be thought of as belonging to every one of the world's citizens. If the material used in those structures is considered, the average person in the world is related to some 55 Mg of highway materials. This is lots of stuff, and its very assemblage involves heavy investments in energy and significant environmental impacts.

The transportation infrastructure also includes a wide variety of "hardware": signage, lighting standards, manhole covers, fencing, gates, guard rails, and so on. These standard products can be evaluated on an environmental basis as could be any product, including the degree to which recycled materials are used in their construction. Among current success stories are the use of recycled plastic in sign posts, plastic drums, cones, and drain pipe, recycled steel in steel reinforcing bars, and recycled glass in reflective pavement markings.

Table 7.4   Embedded Resources in the World's Roadways

Resource	Embedded Amount (Tg)	Embedded Amt. per Capita (kg)
Asphalt Roadways:		
Aggregate	280	50,000
Bituman	4.5	810
Concrete Roadways:		
Aggregate	19	3400
Sand	2.2	390
Cement	0.88	160
Reinforcing Steel	0.0003	0.005

SUGGESTED READING

Ginley, D.M., Material flows in the transport industry, an example of industrial metabolism, *Resources Policy, 20*, 169–181, 1994.

Ginley, D.M., R. Hurdelbrink, and J.F. Lemons, Jr., Intermaterial competition in transportation and construction, *Materials and Society, 13*, 211–232, 1989.

Smith, V.K., Ed., *Scarcity and Growth Revisited*, Johns Hopkins Univ. Press, Baltimore, MD, 298 pp., 1979.

U.S. Bureau of Mines, *Mineral Commodity Summaries 1993*, Washington, D.C., 201 pp., 1993.

EXERCISES

7.1   If the lead in lead-acid batteries is heavily recycled, why have electric cars, which use substantial amounts of lead-acid batteries for energy storage, been criticized on environmental grounds?

7.2   Assume that you work for an automobile manufacturing firm, and have been assigned to engineer "fluff" so that it can be burned for energy recovery. Discuss the environmental costs and benefits of this option. What problems might such an approach generate in automobile design? For the materials, parts, and overall automobile recycling system?

7.3   49 million new vehicles were manufactured in 1990 by the world's auto industry. How much steel was consumed as a result? If the vehicles had used as much steel as they did in 1950, what would have been the result?

7.4   The typical 1990s automobile consumed 65 kg of aluminum. What proportion of this material eventually underwent horizontal recycling? What proportion underwent vertical recycling? What proportion was lost?

7.5   Repeat the previous problem for all materials combined.

CHAPTER 8

Environmental Interactions During Manufacture

"A combination of increasing regulatory pressure, mounting public expectations, and tightening competitive conditions is now driving companies everywhere to adopt the logic of pollution prevention."

— Stephan Schmidheiny, ANOVA Holdings, Switzerland

8.1 MAKING THE AUTOMOBILE, THEN AND NOW

While it is fairly straightforward to picture and describe automobile and infrastructure manufacturing processes, there has historically been considerable development in the types of processes used in that manufacture. As examples, let us examine two extremes: the automobile as it was manufactured in the 1950s and in the 1990s.

The generic auto manufacturing process of the 1950s is shown in Fig. 8.1a. This is a simplified diagram that omits many of the components such as windows, seats, tires, and the like while centering on the steel that comprised all the major parts of the vehicle. The sheet steel was formed into products such as body panels, chassis beams, oil pans, and dashboards. After forming, the metal parts were cleaned and some were welded together to form the frame. The panels and other components were then fastened to the frame and the body painted. Meanwhile, steel ingots were melted and sand cast to produce engine blocks, transmission casings, and other high cross-section components. The molded components were trimmed, smoothed, and cleaned, some were plated, and all were then assembled into the active constituents of the automobile. The entire system was tested and finally delivered to the dealer or customer.

The generic auto manufacturing process of the 1990s is shown in Fig. 8.1b. Essentially the same processes are used for the steel in the automobile, though, as will be seen,

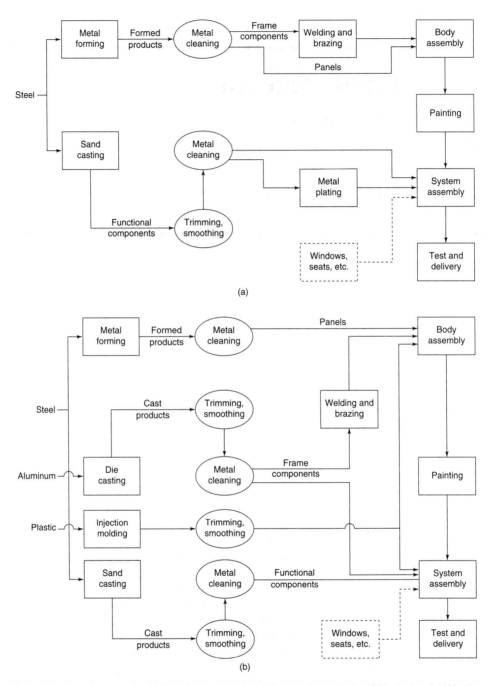

Figure 8.1   Process sequence diagrams for generic automobile manufacture in the 1950s (top) and 1990s (bottom). Primary processes appear in rectangles, complementary processes in ovals. The boxes in dashed lines are included to indicate that only the principal metal structural components and their processes are treated, but that a more comprehensive analysis would include other components and their processes as well.

some aspects of the processes have been significantly improved over the years. The big difference is that aluminum and plastic, virtually nonexistent in the 1950s automobiles, are widely used in 1990s automobiles. Aluminum appears in frames, engines, and engine components, and plastics are used as front ends, non-load bearing body panels, and interior fittings. The aluminum parts are cast in dies made of steel. (This is possible because aluminum melts at much lower temperatures than does steel.) As with sand casting, die casting is followed by trimming, smoothing, and cleaning. A different approach is taken to manufacture the plastic parts, which are injection-molded, trimmed, and smoothed, and can generally be used without further cleaning. The final assembly operation for the 1990s vehicle is essentially similar to that of four decades earlier.

8.2 MANUFACTURING PROCESSES FOR AUTOMOBILES

The most significant processes used in automobile manufacture are sand casting, die casting, metal forming, injection molding, welding and brazing, metal plating, painting, trimming and smoothing, and metal cleaning. The latter two are complementary processes; that is, they do not accomplish a manufacturing purpose in themselves, but are needed to permit the primary process to be performed satisfactorily. The most important characteristics of all these processes from an industrial ecology standpoint are given in Tables 8.1 and 8.2 and discussed below.

8.2.1 Sand Casting

Casting is the process of forming objects by pouring or injecting liquid into a mold. Components of steel that involve detailed shapes and patterns are generally manufactured by sand casting. In this process, selected and prepared sand is packed into a steel mold housing, and the reverse image of the component is formed in sand coated with an organic binder (generally by making an image of the component to be manufactured during casting), forming the mold around the image, and then removing the image. Molten steel is then poured into the mold and allowed to harden. The sand is brushed away from the component, which can then be cleaned and further processed.

Transforming the metal from the solid to the molten state requires substantial energy, much of which is eventually lost as heat. The continuous cycle of heating and cooling also consumes considerable energy. The organic binder on the loose sand has undergone chemical reactions in the casting process, and after casting it contains potentially hazardous materials such as polynuclear aromatic hydrocarbons. As a consequence, and because cleaning the sand is expensive and new sand is cheap, the used foundry sands are discarded as waste after several casting cycles. The flow diagram of materials, energy, and components for sand casting is shown in Fig. 8.2a.

8.2.2 Die Casting

The manufacture of aluminum automotive parts is generally carried out by die casting. In this operation, a die of tool steel forms the reverse image of the desired component. Molten

Table 8.1   Salient Characteristics of Processes for Generic
Automobile Manufacturing of the 1950s and 1990s

Characteristic	ca. 1950s	ca. 1990s
Sand Casting		
Implementation	Ubiquitous	Ubiquitous
Matl. Source	Virgin	Some recycled
Energy Use	Very high	High
Residues	Contaminated sand	Contaminated sand
Die Casting		
Implementation	No	Ubiquitous
Matl. Source	N/A	Some recycled
Energy Use	N/A	High
Residues	N/A	Wastewater
Metal Forming		
Implementation	Ubiquitous	Ubiquitous
Matl. Source	Virgin	Some recycled
Energy Use	Very high	High
Residues	Trimmings, lubricants	Trimmings, lubricants
Welding/Brazing		
Implementation	Ubiquitous	Ubiquitous
Matl. Source	Virgin	Some recycled
Energy Use	High	Moderate
Residues	Negligible	Negligible
Metal Plating		
Implementation	Common	Rare
Matl. Source	Virgin	Some recycled
Energy Use	Moderate	Moderate
Residues	Toxic wastewater	Toxic wastewater
Injection Molding		
Implementation	Nonexistent	Ubiquitous
Matl. Source	None	Some recycled
Energy Use	None	High
Residues	None	VOC
Painting		
Implementation	Ubiquitous	Ubiquitous
Matl. Source	Virgin	Some recycled
Energy Use	Moderate	High
Residues	Much VOC	Minor VOC

Table 8.2   Salient Characteristics of Secondary Processes for Generic
Automotive Manufacturing of the 1950s and 1990s

Characteristic	ca. 1950s	ca. 1990s
Trimming, Smoothing		
Implementation	Ubiquitous	Ubiquitous
Matl. Source	Virgin	Some recycled
Energy Use	Very high	High
Residues	Metal trimmings	Metal trimmings
Metal Cleaning		
Implementation	Ubiquitous	Ubiquitous
Matl. Source	Virgin	Some recycled
Energy Use	Moderate	Moderate
Residues	CFCs	Organics

aluminum is forced under pressure into the die and allowed to harden. An organic "parting
agent" is sprayed onto the die surface prior to pouring, in order to keep the molten aluminum from welding itself to the die. After each part is molded, the die must be cooled
from above $1000°C$ to less than $400°C$ by cold water spray. The spray also cleans degraded
parting agent from the die surface, which is then recoated before the next casting is made.

Transforming the metal from the solid to the molten state requires substantial energy,
much of which is eventually lost as heat (although this is minimized in some cases by the
delivery of molten metal to the automaker from a contiguous foundry). In addition, the
water used to cool the die becomes contaminated by the parting agent, and must undergo a
purification step before reuse; alternatively, it is discarded. The flow diagram of materials,
energy, and components for die casting is shown in Fig. 8.2b.

8.2.3   Metal Forming

Forming is a process by which the size or shape of a metal part is changed by producing
stresses in the part that exceed the yield strength but not the fracture strength. There are
many varieties of forming, involving pressing, hammering, rolling, and drawing; most are
used in the manufacture of one or another component of the typical automobile, such as
body panels and engine, transmission, and suspension parts. The metal forming equipment
that contacts the parts is of tool steel, while the remainder of the equipment is generally of
carbon steel.

Forming processes are energy-intensive, and significant heat is produced as a
by-product. Lubricants are often utilized to enhance machinability, thus requiring the
purification or disposal of contaminated lubricants. Additionally, most forming processes
are followed by a trimming stage to remove excess metal; the trimmed material must then

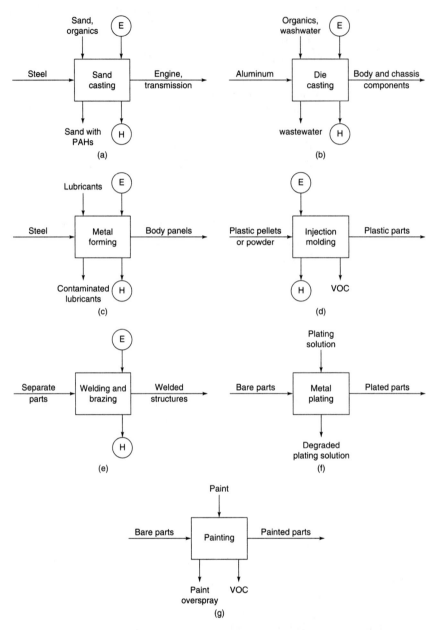

Figure 8.2   Representations of flows of materials, energy, and components for primary processes used in auto-mobile manufacture. (a) Sand casting; (b) Die casting; (c) Metal forming; (d) Injection molding; (e) Welding and brazing; (f) Metal plating; (g) Painting. Components enter the process from the left and leave from the right, except that occasionally this format is modified to improve clarity. Process materials enter from the top and waste products leave from the bottom. Heavy lines indicate flows of products, lighter lines flows of process resources. A circled E indicates energy input, a circled H indicates heat output. VOC referes to volatile organic carbon gases. The list of processes is intended to be illustrative, not comprehensive.

be recycled in some fashion. The flow diagram of materials, energy, and components for metal forming is shown in Fig. 8.2c.

8.2.4   Injection Molding

Plastic is received from the chemical manufacturer in the form of pellets perhaps a centimeter in size, or as a powder. The pellet or powder material is fed into a molding machine, in which as a screw drive moving through a heated cylinder heats it until it melts ("plasticizes") into a viscous liquid. Periodically, the nozzle at the end of the cylinder is thrust against an opening to the mold and the liquid is injected under high pressure into the mold cavity. The molded plastic cools and hardens to form a part in the shape of the cavity, the mold is opened, and the part removed.

The injection molding operation is energy-intensive, since both the heating and injection operations require substantial energy inputs. Volatile organic gases are produced during the heating process; these must generally be captured rather than released to the ambient environment. Mold scrap is produced in fairly large amounts, but can generally be recycled back into the mold as a portion of the incoming feed stream. The flow diagram of materials, energy, and components is shown in Fig. 8.2d.

8.2.5   Welding and Brazing

The process of welding or brazing is one in which metal parts are joined by applying heat, and sometimes pressure, at the joint between the parts. In spot welding, the dominant technique, no added metal is required. In some instances, however, a metal is added to fill in the joint being melted. In electric-arc and gas-torch welding, the filler metal is the same as the base metal. In brazing and soldering, the filler metal (which has a lower melting point) is melted, but the base metal is not.

Welding requires substantial amounts of energy to melt the filler metal (and often the base metal). No residues of consequence are produced. The flow diagram of materials, energy, and components for welding and brazing is shown in Fig. 8.2e.

8.2.6   Metal Plating

Plating is the process of depositing a thin layer of one metal on another, generally for protective and/or decorative purposes. In electroplating, the piece to be plated is immersed in a solution and made the cathode of a direct current circuit. The coating metal acts as the anode, replenishing the solution as its ions are attracted to the piece being plated. In electroless plating, the metal is deposited from solution without an imposed current being applied. Plating tanks are generally of carbon steel, and may be lined with an organic corrosion-protection layer.

Most plating solutions and most plating metals that are highly protective, such as chromium, are also highly toxic. As a consequence, metal plating has rapidly decreased in frequency of use in recent years, and parts that formerly were of plated metal are often replaced with parts formed from plastics or composites. The flow diagram of materials, energy, and components for metal plating is shown in Fig. 8.2f.

8.2.7   Painting

The painting operation (or, more generally, the surface coating operation) is performed on essentially all visible exterior and interior body components on all products. Painting involves the application of the pigment in a solvent carrier. The carrier has traditionally been organic in nature, though some current carriers are aqueous solvents, and the industry is now beginning to utilize solvent-free electrostatic powder painting techniques.

The painting operation involves the capture and disposal of paint overspray and the emission of greater or lesser quantities of volatilized paint and carrier. Because of the large quantities of air used in modern painting systems, the energy consumed is large. The flow diagram of materials, energy, and components for painting is shown in Fig. 8.2g.

8.2.8   Trimming and Smoothing

The use of a casting or injection molding process requires the use of a complementary process, trimming and smoothing, as well. After castings or moldings are removed from the molds or dies in which they are made, there is invariably excess metal or plastic (termed "sprues" and "runners") that results from mold seams, pouring spouts, and the like. This excess material must be trimmed away and the metal or plastic then smoothed so that the product will fit properly and have a satisfactory appearance.

Trimming and smoothing are not energy-intensive, but normally involve the use of tool-steel machinery to cut and abrade. Organic lubricants are commonly used. The flow diagram of materials, energy, and components for trimming and smoothing is shown in Fig. 8.3a.

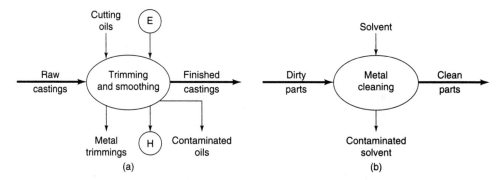

Figure 8.3   Representations of flows of materials, energy, and components for complementary processes used in automobile manufacture. (a) Trimming and smoothing of castings and moldings (cutting oils are used on metals but not plastics); (b) Metal cleaning. Symbols and diagrammatic protocols are as in Figure 8.2. The list of processes is intended to be illustrative, not comprehensive.

8.2.9   Metal Cleaning

A second complementary process in automobile manufacture is metal cleaning. During casting, forming, and other process operations, metal parts tend to collect dirt, oil, and other contaminants, which must be removed prior to surface coating or other contaminant-sensitive operations. The cleaning step generally involves the use of liquid solvents, organic and inorganic, in a lined steel tank, often followed by an aqueous rinsing stage.

For many years, chlorofluorocarbons (CFCs) were the solvent of choice for metal cleaning as a consequence of their nonflammability, nontoxicity, and satisfactory cleaning ability. With the discovery of the effects of CFCs on the stratospheric ozone layer, CFCs have generally been replaced by hydrogenated variants (HCFCs) or other alternatives. No matter what the cleaning solvent, contaminated solvent is the inevitable result of metal cleaning processes; the solvent must then be purified in order to be reused. Alternatively, it can be disposed of in a suitable manner. The flow diagram of materials, energy, and components for metal cleaning is shown in Fig. 8.3b.

8.3   GREENING THE AUTOMOBILE MANUFACTURING PROCESS

Many initiatives are underway to make automobile manufacturing more environmentally responsible. One that has achieved great gains has been the reduction of energy use, produced largely by attention to energy minimization in process equipment design. Better control of molds has reduced the volume of metal trimmings, but not eliminated them. Some reuse of wastewater is being accomplished in mold cleaning and water treatment operations.

Despite this progress, much remains to be done. The most advanced industrial corporations are beginning to adopt "zero emissions" as a goal, but most manufacturing facilities in the automobile industry and elsewhere are far from accomplishing this task. Making substantial further progress may involve a transition to new processes rather than merely improvement of the old. Metal casting, for example, will probably always use large amounts of energy and produce large volumes of sand or wastewater. Alternative approaches for forming metals, plastics, and composites are continuously being implemented to minimize the overall system's impact on the environment.

8.4   MANUFACTURING PROCESSES FOR THE AUTOMOTIVE INFRASTRUCTURE

The most significant processes used in automobile infrastructure manufacture are minerals manufacture, asphalt pavement manufacture, cement manufacture, metals manufacture, and concrete pavement manufacture. Their most important characteristics from an industrial ecology standpoint are given in Table 8.3 and discussed below.

Table 8.3   Salient Characteristics of Processes for Automotive
Infrastructure Manufacturing of the 1950s and 1990s

Characteristic	ca. 1950s	ca. 1990s
Minerals Mfr.		
Implementation	Ubiquitous	Ubiquitous
Matl. Source	Virgin	Mostly virgin
Energy Use	Very high	Very high
Residues	Rock dust	Rock dust
Asphalt Pvt. Mfr.		
Implementation	Ubiquitous	Very common
Matl. Source	Virgin	Much recycled
Energy Use	Very high	Very high
Residues	Solids, VOC	Some solids, VOC
Cement Mfr.		
Implementation	Modest	Common
Matl. Source	Virgin	Some recycled
Energy Use	Very high	Very high
Residues	CO_2, water, solids	CO_2, water, solids
Metals Mfr.		
Implementation	Ubiquitous	Ubiquitous
Matl. Source	Virgin	Some recycled
Energy Use	Very high	Very high
Residues	SO_2, slag, water	SO_2, slag, water
Concrete Pvt. Mfr.		
Implementation	Modest	Common
Matl. Source	Virgin	Mostly virgin
Energy Use	Very high	Very high
Residues	Some solids	Some solids

8.4.1   Site Preparation

Large components of the automotive infrastructure such as roadways and concrete overpasses are manufactured on-site rather than elsewhere, and site preparation often involves
the removal or movement of substantial amounts of materials; rock, soil, trees, sometimes
existing buildings. This step is, strictly speaking, not infrastructure manufacture but preparation for manufacture. It is, nonetheless, a step that often involves major impacts on the
environment, and is included here so that it is not overlooked. A roadway manufacturer
who intends to be environmentally responsible should exercise due diligence at the site
preparation step just as the auto body manufacturer should operate an environmentally
responsible factory. Aspects of site preparation that should be considered include

- To the extent possible, preserve and/or restore local ecosystems
- Minimize materials waste during demolition activities
- Recycle material extracted or moved during site development
- Recycle material in existing buildings that are to be demolished
- Select site preparation materials that themselves have low environmental impacts
- Use energy-efficient equipment in site development
- Reduce the use of water in site development
- Reduce the potential for, and volume of, contaminated runoff.

8.4.2   Minerals Manufacture

Minerals (broadly interpreted to include rock, sand, and stone) are the largest components by weight of the automotive infrastructure. Their manufacture is conceptually simple, though logistically demanding. As shown in Fig. 8.4a, it begins with the extraction of the material from the mineral body, and generally proceeds with crushing and perhaps grinding to produce fragments a few centimeters or less in size. Size segregation is then performed by some form of sieving. Often this material can be used directly, as in the aggregate used to make asphalt and concrete. Alternatively, it may undergo any of a variety of forms of purification if a pure mineral is required, as in the use of lime for soil stabilization.

8.4.3   Asphalt Pavement Manufacture

Asphalt pavement is manufactured on site by mixing bitumen, stone, and mineral dust in the proper proportions (Fig. 8.4b). To begin, the bitumen is heated to $150°C$ to render it semi-liquid. The stone and mineral dust (the "aggregate") are then added and mixed, and the mixture spread on the pavement base. Most mixtures harden within a few hours. The byproducts of the process are heat and a minor amount of solid residues. It is worth noting that not all asphalt systems are the same; some last much longer than others, providing a form of "product life extension" and reducing the demand for maintenance material and early replacement. The use of "lowest cost" bidding procedures, which emphasize initial cost over full life-cycle cost, unfortunately favor the less robust alternative.

8.4.4   Cement Manufacture

The manufacture of cement is a very large activity throughout the world, as it is the basis for modern buildings and infrastructure. One begins by combining limestone ($CaCO_3$), shale (a composite mineral of Al_2O_3, SiO_2, and Fe_2O_3), and water, and then heating the mixture to $1300°C$ in a rotating kiln (Fig. 8.4c). The resulting "clinker" material consists of calcium silicates and calcium aluminates. The clinker is ground to a powder and combined with gypsum to produce the cement. Typical proportions of the ingredients are 1000 kg of shale, 4800 kg of limestone, 140 kg of gypsum, and 6200 l of water.

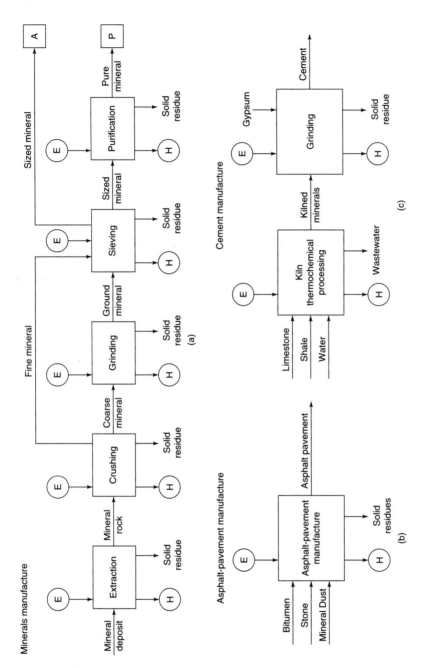

Figure 8.4 Representations of flows of materials, energy, and components for primary processes used in automotive infrastructure manufacture. (a) Minerals manufacture; (b) Asphalt pavement manufacture; (c) Cement manufacture; (d) Metals manufacture; (e) Concrete pavement manufacture. In 8.4a, output A indicates a mineral material that will be used because of its physical properties, as with aggregate in making asphalt. Output P indicates a mineral material that will be used because of its chemical properties, as with gypsum in making cement. In 8.4d only the major process materials and waste products are indicated. Other symbols and diagrammatic protocols are as in Figure 8.2. The list of processes is intended to be illustrative, not comprehensive.

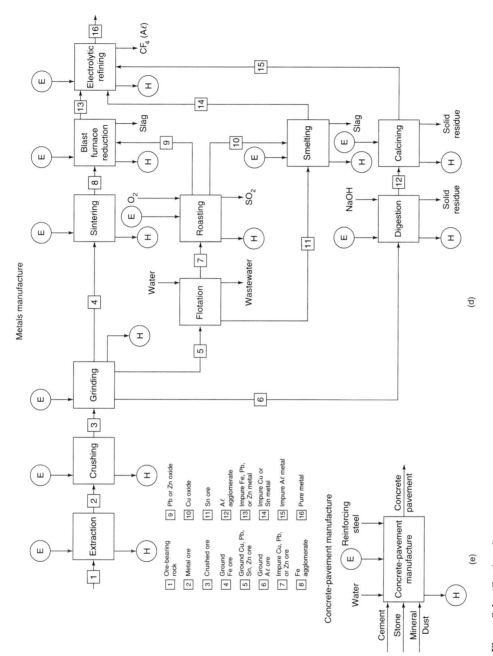

Metals manufacture

1	Ore-bearing rock	
2	Metal ore	
3	Crushed ore	
4	Ground Fe ore	
5	Ground Cu, Pb, Sn, Zn ore	
6	Ground Aℓ ore	
7	Impure Cu, Pb, or Zn ore	
8	Fe agglomerate	
9	Pb or Zn oxide	
10	Cu oxide	
11	Sn ore	
12	Aℓ agglomerate	
13	Impure Fe, Pb, or Zn metal	
14	Impure Cu or Sn metal	
15	Impure Aℓ metal	
16	Pure metal	

(d)

Concrete-pavement manufacture

(e)

Figure 8.4   (Continued)

105

8.4.5   Metals Manufacture

The extraction and refining of metals is a large subject in itself, and one we can treat here only in outline. Our approach emphasizes not the engineering details of the processes themselves, but the flows of resources, byproducts, and energy through the processes.

No matter what the metal, the initial steps in the manufacturing cycle are similar (Fig. 8.4d). The ore-bearing rock is extracted, crushed, and ground into material suitable for further handling. The next step depends on the metal that is to be recovered. In the case of iron, present in the ore as Fe_2O_3, the ground ore is sintered, typically at temperatures over $1000°C$, to form an agglomerate of concentrated ore. The agglomerate is chemically reduced in a blast furnace, and electrolytically refined to produce the usable metal.

If aluminum is the metal being produced, the ore grinding is followed by digestion, in which sodium hydroxide, heat, and pressure separate the alumina (Al_2O_3) from various impurities in the ore. A calcination step (heating the mineral to below its melting point) is then performed to dissociate and remove the combined water. The relatively pure alumina is then dissolved in molten cryolite (Na_3AlF_6) and electrolyzed to produce the pure aluminum metal.

The procedure for the major industrial metals copper (Cu), lead (Pb), tin (Sn), and zinc (Zn) is more complex because those metals usually occur in combination. The processing must thus include separation, usually accomplished initially by flotation, in which the surface tension of ore particles of specific compositions are altered by adding a variety of chemical compounds. Air is then added to adhere to the altered particles, which are carried to the liquid surface and removed.

An additional complication is that copper, lead, and zinc are usually found in nature as sulfides, not oxides. This requires that they be converted to oxides by the addition of oxygen under heat, a process known as roasting. This is followed by smelting, in which the mixed metals are recovered sequentially under high heat. The separated metals may then be refined electrolytically.

The environmental consequences of metals manufacture are extensive:

- For all metals, the amount of rock that must be extracted to recover the desired ore is very large. In the cases of copper and aluminum, rich ores can still be found, but for most metals the ores have grades much below 10%, so that more than 90% of the rock that is moved in the extraction process is discarded.
- The use of energy in almost all stages of metal processing is large. As a consequence, the emissions of the greenhouse gas CO_2 are very high on a weight basis.
- The use of water in several stages of metals manufacturing is large, as is contamination of the water unless control and treatment facilities are employed. Water degradation occurs not only during the active life of a mine, but can be a problem even after the mine is abandoned.
- The processing of copper, lead, and zinc produces substantial amounts of sulfur dioxide gas. This byproduct can be, and increasingly is being, captured; if not, it can be a major precursor to acid rain.

• The processing of aluminum results in the emission of CF_4, a greenhouse gas. Modern processing methods have strongly decreased the rate of emission of this byproduct.

8.4.6   Concrete Pavement Manufacture

The manufacture of concrete pavement generally (though not always) begins with the placement of reinforcing steel on the pavement base layer. Cement, stone, mineral dust, and water, in precise proportions, are then mixed together (Fig. 8.4e). When water is added to the cement, the calcium silicates in the cement are largely transformed into tobermorite gel (an amorphous, hydrated, calcium-containing product) and crystalline calcium hydroxide. The tobermorite gel, about 50% by weight of the reacted cement, is the principal adhesive agent joining the aggregate together. The mixture is poured into place, where it hardens within a few hours, though complete curing takes days to weeks.

8.5   GREENING THE AUTOMOTIVE INFRASTRUCTURE MANUFACTURING PROCESS

Greening automotive infrastructure manufacturing is a difficult prospect because infrastructure involves such great quantities of materials (and thus great quantities of energy). Intelligent site preparation will help, however, as will increased attention to the recycling of such construction materials as wooden forms, asphalt residues, and component packaging. The use of water and the production of wastewater in site development, cement manufacture, and roadway and bridge manufacture can probably be sharply reduced.

As with the automobile itself, materials substitution and dematerialization in the infrastructure should receive increased attention. A variety of materials are now being used as additives for asphalt and concrete, and it is possible that the total materials volume in the infrastructure may decrease as thinner, more rugged materials and techniques are developed. Infrastructure design and development engineers need to consider the advent of modern tailored materials as an opportunity to modify or change road building, one of the earliest and one of the least-changed of the engineering specialties.

SUGGESTED READING

Bodsworth, C., *The Extraction and Refining of Metals*, Boca Raton, FL: CRC Press, 348 pp., 1994.

Kalpakjian, S., *Manufacturing Engineering and Technology*, 2nd ed., Reading, MA: Addison-Wesley, 1992.

Krieger, J.H., Zero emissions gathers force as global environmental concept, *Chemical and Engineering News, 74* (28), 8–16, 1996.

Shreve, R.N., *Chemical Process Industries*, 2nd ed., New York: McGraw-Hill, 1004 pp., 1956.

Wills, B.A., *Mineral Processing Technology*, 3rd ed., Oxford, U.K.: Pergamon Press, 629 pp., 1985.

EXERCISES

8.1 What properties of materials commonly used in automobiles are most important at the extraction/initial production stage? At the manufacturing stage? At the in-use stage? At the end-of-life stage?

8.2 Energy consumption has often been expressed in quads, a quad being 10^{15} BTU. In 1989 the world's energy consumption was 385 quads. Express the global energy consumption by human activity in exajoules (EJ).

8.3 The total U.S. electrical energy use in 1985 was approximately 9.4 EJ. If the energy equivalents of a metric ton of coal and a 160 liter barrel of oil are 29.6 GJ and 610 GJ respectively, find how many metric tons of coal would have been required to produce the electrical energy if coal combustion was the sole source? How many barrels of oil would have been required if oil combustion was the sole source?

8.4 In 1990, the burning of fossil fuels resulted in the emission of about 5.5 Pg C to the atmosphere as CO_2. The total CO_2 concentration in the atmosphere was 750 Pg C. If 45% of the emitted CO_2 remained in the atmosphere, by what percentage did the atmospheric concentration increase? How many years of such emissions would it take to double the atmospheric CO_2 concentration?

8.5 Although the proportion is uncertain, it is estimated that about 45% of the CO_2 currently emitted as a result of human activity remains in the atmosphere. If the 1990 CO_2 concentration in the lower atmosphere was 350 ppm (parts per million by volume), what CO_2 concentration do you predict by 2020 if emissions grow by 4% (uncompounded) per year?

8.6 If the average vehicle is 4 years old, travels 10,000 km per year, and has the fuel economy given in Fig. 5.7, compute the fuel consumption of the world's vehicles in 1979 and 1992, using the fleet sizes of Fig. 5.2.

CHAPTER 9

Energy Consumption

"Society needs to take out some technological insurance against the possibility that the world will have to curtail fossil fuel consumption drastically and rapidly to prevent global climate change."

— William Fulkerson, Oak Ridge National Laboratory

9.1 THE PETROLEUM INFRASTRUCTURE

The modern automobile is dependent on petroleum in many ways, and the analysis of the automobile and its supporting infrastructure thus requires a perspective on petroleum extraction, transport, refining, and delivery to customers. Aspects of this system are shown schematically in Fig. 9.1. The process begins with the drilling and development of a well to extract the petroleum from its reservoir. The crude oil that is recovered may be subsequently pumped directly to a nearby refinery or it may be transported by seagoing tanker to a distant refinery.

Most of the oil pumped from Earth's underground reservoirs is not used in the region where it is extracted, but rather is shipped, mostly by sea, to other regions. A picture of that exchange for a typical recent year is shown in Fig. 9.2. As is well known, the Middle East is the leading producer and exporter, followed by Africa, South America, and the Former Soviet Union. Europe is the leading importer, followed by North America and Japan.

At the refinery the crude oil is heated and then distilled into a number of components, the most significant from the automotive standpoint being shown in Fig. 9.1. These products of the petroleum refining activity are then themselves shipped either to the transportation infrastructure manufacturing network (the bitumen) or to the vehicle fuel distribution network (gasoline and diesel fuel). All in all, something more than half of the world's petroleum used in each year is consumed in service to automobiles and their infrastructure.

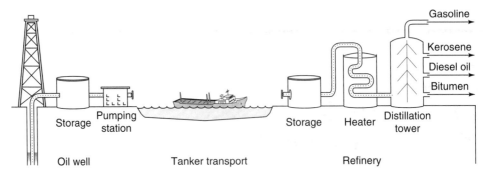

Figure 9.1   A schematic diagram of the extraction and processing of petroleum.

Petroleum is the primary feedstock for the manufacture of the world's plastics and many of its other chemicals, and no ready substitute exists for most of these processes. Given the importance of the petroleum supply to those industries as well as to the transportation system, it is of interest to assess the long-term supplies of petroleum. How much oil is left in the ground to be recovered and used? The question cannot be answered with precision, since new deposits continue to be discovered and new methods continue to be developed to recover oil from reservoirs difficult to reach. Statistics are available, however, for "proven reserves," the amount of petroleum known to be present in specific locations. Figure 9.3 illustrates, for different regions of the world, both the proven reserves as of 1989

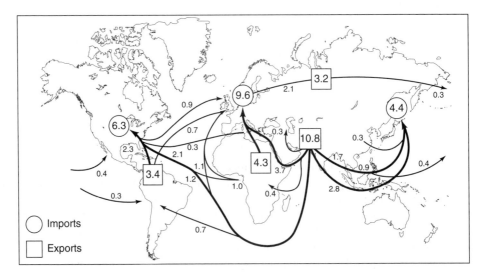

Figure 9.2   The movement of oil by sea. The data are for 1987, in units of millions of barrels per day. (Reproduced by permission from M.K. Tolba et al., Eds., *The World Environment, 1972–1992, Two Decades of Challenge*, London: Chapman and Hall, on behalf of the United Nations Environment Programme.)

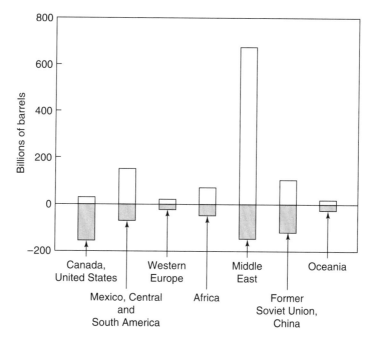

Figure 9.3   Recovered and proven reserves of petroleum in different global regions. Positive numbers are reserves, negative numbers are the total amounts of oil recovered from the regions between 1859 and the present. (Adapted from a diagram devised by the United States Office of Technology Assessment.)

and the total petroleum already extracted from those reservoirs during the time period 1859–1989. At the present time, the rate of use of petroleum is estimated to be some 15,000 times the rate at which new petroleum is being formed, a balance between use and replenishment that is clearly unsustainable over the long term. Even though one must acknowledge that the reserves may be augmented somewhat by new discoveries in the future, the general picture will not change. The potential market is dominated by the Middle East, which has 65 percent of the proven reserves. Most of the rest is in Central and South America, in a few countries of the former Soviet Union, and in Africa.

9.2   ENERGY CONSUMPTION IN AUTOMOBILE MANUFACTURE

The industrial sequences involved in the manufacture of automobiles are sufficiently diverse and technologically advanced that they provide a microcosm of the entire industrial process. The life cycle sequence for an automobile, with emphasis on the manufacturing stage, is shown in Fig. 9.4. Notice that "energy" or "fuel" appear nearly everywhere on the diagram. Aspects of this diagram will be discussed throughout this book, but for the present it is

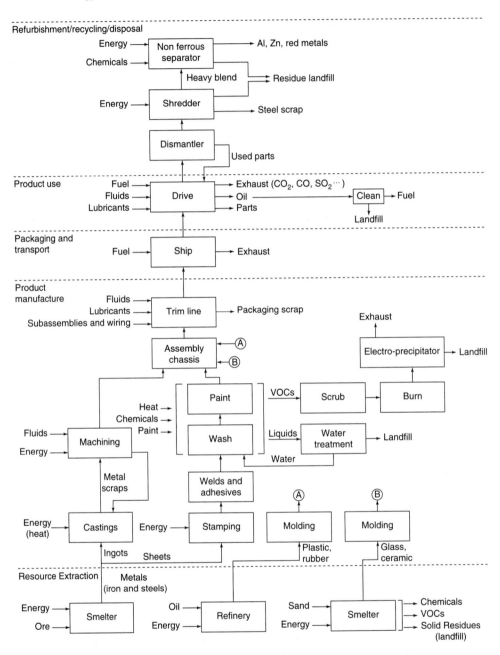

Figure 9.4   The life cycle of a typical automobile. The life cycle flow is from bottom to top. Materials and energy inputs enter from the left, residues leave to the right.

appropriate to comment on major uses of energy in resource extraction and product manufacture.

Resource extraction is typically an energy intensive life stage, and the smelting and refining that comprise automotive raw material processing are good examples. The largest inputs are ores for iron, steel, and aluminum. Petroleum is the feedstock commonly used to produce the plastics and rubbers so widely used in vehicles, and sand is used as the precursor to glass and ceramics.

Within the production process itself, the largest energy use activities are those dealing with high temperature or high pressure processes: casting, stamping, welding, and machining. Energy utilization in these processes has been improved considerably over the years, but much room for improvement still exists.

Good energy housekeeping in the manufacturing facility involves taking the industrial situation as it exists and devising ways to modify or change it to make it more energy efficient. This activity can be aided by checklists and lists of relevant questions. We suggest the following as aids to energy analysis.

Process Designers:

- Is the process designed with the aim of minimizing the use of energy-intensive process steps such as high heating differentials, heavy motors, extensive cooling, cycling between warm and cool states, etc.?
- Is the process designed to optimize the use of heat exchangers and similar devices to utilize otherwise wasted heat?
- Does the process use the maximum possible amount of recycled material rather than virgin material? (To some extent, the appropriateness of this recommendation depends on materials choice: for aluminum, recycling saves significant amounts of energy; the opposite is true of glass.)
- Is the process designed to utilize energy management approaches and equipment to minimize energy use?
- Is the process designed to utilize variable speed motors and other automated load controls?
- Have energy-reduction targets for research and development activities been identified? (For example, the auto industry is actively seeking alternatives to die casting that reduce or eliminate the heat/cool cycle now necessary for each casting.)

Facility Engineers:

- Has incandescent lighting been replaced with high-efficiency fluorescent lighting?
- Has an automatic lighting control system been installed?
- Have boilers and furnaces been checked for leaks and repaired as necessary?
- Are boilers correctly sized?
- Is the facility designed to utilize cogenerated heat and electricity from within the facility or nearby?

- If waste heat is produced, has economic use of this resource been explored?
- Have ceilings, walls, and pipes been insulated?
- Does the industrial facility have a program to encourage good energy housekeeping?

9.3 IN-SERVICE ENERGY CONSUMPTION OF THE INDIVIDUAL AUTOMOBILE

Some products—nails, furniture, books, to name but a few—consume no energy when in use. Others, including automobiles, depend on energy for their operation. And unlike washing machines and desk lamps, which draw energy from a network supply when needed, the automobile must carry its energy supply with it, using up the supply as it travels.

Except for a few experimental vehicles, automobiles derive their energy from petroleum. Although average annual distance driven and fuel efficiency differ greatly among users and countries, the global averages are about 8000 km/yr and 10 km/l.

In today's typical midsize sedan, only about 12% of the energy produced by the engine is translated into usable kinetic energy and eventually dissipated as aerodynamic and rolling friction and in braking. As seen in Figure 9.5, engine and drivetrain losses constitute the remainder. Engine losses alone amount to more than 60% for urban driving, 17% is sacrificed to standby losses, and nearly 6% of the energy disappears as driveline loss. Accessories take a small toll of the remainder. Engineers face continuing challenges in improving vehicle utilization of energy while not adversely influencing other environmental

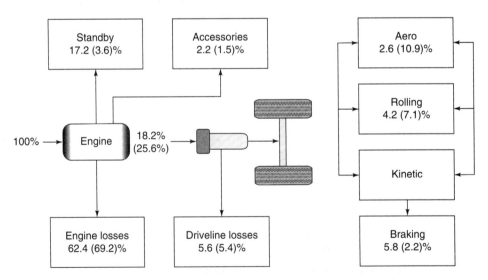

Figure 9.5 The distribution of input energy for a midsize automobile. The numbers within parentheses are for highway driving; those without for urban driving. Kinetic energy is completely lost in braking. (Reproduced with permission from D.F. Cioffi, Physicists find challenges in building better cars, *Physics Today, 48*, 73–75, 1995.)

characteristics such as recycling economics (which currently depend on vehicles having a high steel content).

What are the prospects for reducing automotive energy use? First, vehicles are increasingly getting lighter. Progress depends strongly on vehicle type and design approach, but perhaps a 10% decrease in average vehicle weight is possible by 2010. Second, we can continue to expect efficiency improvements. Estimates differ, but a 10% increase in efficiency appears possible by 2010, and the Rocky Mountain Institute has estimated that a "hypercar" utilizing ultralight construction could do considerably better. Offsetting these factors is a tendency for vehicle owners to continually travel greater distances. A 5–10% increase in average annual vehicle distance travelled seems a reasonable prospect by 2010. Finally, the most difficult trend of all to combat: every year there are more vehicles. Projections suggest that a 30% increase in the world's vehicle fleet may occur by 2010. All things considered, it will take the best engineering we can manage just to keep automotive energy use from increasing from present day amounts.

9.4 GLOBAL AUTOMOTIVE ENERGY CONSUMPTION

We noted earlier that the global vehicle fleet as of 1990 was approximately 550 million vehicles. Since each vehicle consumes some 800 l of gasoline per year, the annual fleet consumption is of order 4×10^{11} l, about half of all global petroleum use. By year 2030, it has been predicted that the fleet may double in size; even with substantial improvements in efficiency, this makes it unlikely that any substantial declines in petroleum consumption will occur.

Petroleum is the global materials resource of most pressing concern for two reasons. First, its use releases carbon dioxide from previously sequestered carbon, and thus increases the potential for climate change. Second, one can foresee the eventual exhaustion of the petroleum supply. The United States, Canada, and Western Europe have reserves that are likely to be essentially depleted within one or two decades at present rates of use. The depletion of the other reservoirs will take longer and will be strongly dependent on the rates of industrial development in Africa, Eastern Europe, and South America as well as on the degree to which technological development can decrease energy demand. Nonetheless, it seems very likely that by the end of the 21st century, no matter how world development plays out, petroleum reserves will be close to exhaustion and the cost of petroleum products will be very much higher than is the case today. It is not implausible that large numbers of vehicles will soon be chasing an increasingly rare fuel supply if no significant fuel-switching takes place. Alternatively, it is possible that gasoline-like hydrocarbons produced by biomass conversion, based on genetically-engineered organisms, can begin to meet some of this demand.

9.5 AUTOMOBILE ENERGY OPTIONS

9.5.1 Alternative Carbon-Based Fuels

A number of carbon-based fuels other than petroleum have been suggested as options for providing energy to vehicles. One of these is methanol, which can be readily produced from natural gas and, with added energy expenditure, from coal or biomass. At least in the short term, the source is a positive attribute, since North America and Europe are better supplied with natural gas relative to their needs than is the case for petroleum. China, a major developing market, has low reserves of natural gas, however. In the longer term, natural gas is anticipated to be reliably available for only about a century, so a new major use of natural gas is perhaps not very attractive. Additionally, computer model analyses suggest that the use of methanol would slightly increase the incidence and severity of photochemical smog. Finally, 100% methanol could not be used given today's vehicles and infrastructure: pure methanol fuel is difficult to ignite, and a manufacturing and distribution infrastructure is not in place.

A second potential alternate carbon-based fuel is compressed natural gas (CNG). CNG is fairly widely used as a vehicle fuel in some parts of the world, especially those with substantial natural gas resources. The performance of CNG-fueled vehicles is satisfactory, but wide use of CNG for vehicle propulsion is unlikely because of several undesirable characteristics. First, emissions of oxides of nitrogen and methane, as well as carbon dioxide, indicate that no significant improvements with respect to smog generation or global warming impact can be expected. Second, accomodating the bulky storage tanks in small passenger vehicles will be difficult. Third, an extensive refueling network will have to be developed. Fourth, countries without internal natural gas resources will have to import the fuel, a process little different from current importing of petroleum. All things considered, it appears unlikely that CNG can have more than a minor influence on energy provisioning for vehicles.

The third potential carbon-based alternative fuel is ethanol, mostly derived from corn or sugar cane. Its use has been proposed partly to lessen dependence on imported vehicle fuels. Ethanol performs satisfactorily as a fuel, either in the pure form or mixed with gasoline. However, its use is likely to increase smog formation and have little significant impact on global warming-related emissions. Moreover, the world's corn production, if used solely for motor vehicle fuel, would satisfy less than 10% of the need. Ironically, the current methods of producing ethanol from corn are, themselves, quite energy intensive, making this option less desirable from an industrial ecology point of view.

The benefits and disadvantages of alternative carbon-based fuels are illustrated in Table 9.1. With the exception of "national security," i.e., avoidance of the need to extensively import vehicle fuels, the alternative fuels offer little in the way of advantages and require substantial infrastructure to make them viable. It is unrealistic to expect them to be significant suppliers of energy for the world's vehicles.

Table 9.1   Potential Benefits of Alternative Carbon-Based Fuels*

Fuel	Cost	Smog	Global Warming	Balance of Trade	National Security
Methanol	No	No	No	No	Yes
CNG	No	No	No	No	Yes
Ethanol	No	No	Bio	Yes	Yes

* Yes means that this attribute of the fuel is superior to gasoline; No means it is not; Bio means that it is superior if produced from biomass, not superior if produced from fossil fuel sources. This table is adapted from J.J. MacKenzie, *The Keys to the Car*, Washington, D.C.: World Resources Institute, 1994.

9.5.2   Electric Vehicles

An electric vehicle is one in which the motive power is provided by a stored energy reservoir (a battery) rather than by the consumption of fuel (as with gasoline or ethanol). Electric vehicles are widely regarded as having the potential to be good alternatives to vehicles powered by gasoline, at least for some specialized uses. The key, as all agree, will be the performance of the batteries. Two types are under active development: the traditional electrochemical battery (ECB) and the electromechanical battery (EMB), generally realized as a flywheel. We describe the batteries and their prospects below.

9.5.2.1   Electrochemical Batteries.   Electrochemical batteries have energy densities of only about one-third of one percent that of gasoline, and vehicle weight increases substantially because of the mass of batteries required. As a result, electric vehicles typically have decreased performance and decreased range, both of which are significant disadvantages from the standpoint of the consumer. Typical acceleration is half to two-thirds the gasoline equivalent, and typical range is about 120 km between recharges. The latter is adequate for typical driving of most people, but is a limitation for occasional longer trip use, particularly since battery recharging requires several hours.

As of 1996, ECB vehicles became commercially available. Ongoing ECB development is vigorous, and many candidate batteries are being researched. Their current attributes are given in Table 9.2, where they are compared with the goals of the U.S. Automobile Battery Consortium. It is apparent that all the ECBs fall short of the long-term goals and mostly of the mid-term goals.

The environmental impacts of ECB vehicles are matters of some dispute. Since the batteries must be recharged by electricity generated in power plants, electric vehicles have been characterized by some as "emission moving" vehicles rather than "zero emission" vehicles. That characterization is not very true for the few electric vehicles in France, where most of the electricity is generated by nuclear power. (Of course, one must take account of the need to deal with the spent nuclear fuel.) Most studies of fossil-fuel-fired power plants for electricity generation suggest that they would produce the needed power while generating somewhat less smog-forming gases (e.g., volatile organic hydrocarbons) but more acid-forming gases (e.g., sulfur dioxide).

Table 9.2   U.S. Automobile Battery Consortium Technology Goals and Performance Characteristics of Current Batteries (1994)

	Mid-Term	Long-Term	Present Status
Specific Power (W/kg)	150	400	
Lead-acid			67–138
Nickel-iron			70–132
Nickel-cadmium			100–200
Nickel-metal hydride			200
Sodium sulfur			90–130
Sodium nickel chloride			150
Lithium polymer			100
Energy Density (Wh/L)	135	300	
Lead-acid			50–82
Nickel-iron			60–115
Nickel-cadmium			60–115
Nickel-metal hydride			152–215
Sodium sulfur			76–120
Sodium nickel chloride			160
Lithium polymer			100–120
Specific Energy (Wh/kg)	80	200	
Lead-acid			18–56
Nickel-iron			39–70
Nickel-cadmium			33–70
Nickel-metal hydride			54–80
Sodium sulfur			80–140
Sodium nickel chloride			100
Lithium polymer			150
Life (years)	5	10	
Lead-acid			2–3
Nickel-iron			
Nickel-cadmium			
Nickel-metal hydride			10
Sodium sulfur			
Sodium nickel chloride			5
Lithium polymer			
Cycle Life (80% DOD)	600	1000	
Lead-acid			450–1000
Nickel-iron			440–2000
Nickel-cadmium			1500–2000
Nickel-metal hydride			1000
Sodium sulfur			250–600
Sodium nickel chloride			600
Lithium polymer			300

Table 9.2   (Continued)

	Mid-Term	Long-Term	Present Status
Ultimate Cost (dollar/kWh)	<150	<100	
Lead-acid			70–100
Nickel-iron			160–300
Nickel-cadmium			300
Nickel-metal hydride			200
Sodium sulfur			100+
Sodium nickel chloride			>350
Lithium polymer			50–500

Another consideration is the efficiency with which used ECBs would be recycled. In the case of lead-acid batteries, the most likely at present due to cost, performance, and existing manufacturing capability, recycling losses of lead occur during secondary smelting and remanufacturing. There is thus the possibility that the health consequences of such losses may exceed the air quality benefits of battery operation.

9.5.2.2   Electromechanical Batteries.   An alternative to the ECB is the electromechanical battery, in which energy is stored in a flywheel rotating rapidly in a vacuum. The specific power of the EMB is substantially greater than the ECB, and EMBs have the potential to be maintenance-free for perhaps a decade. EMBs have principally been evaluated in so-called "hybrid vehicles", discussed in the next section, where, unlike the ECBs, they could be recharged by capturing the energy that would otherwise be lost by braking (i.e., their rotation speed, which had been reduced by loss to vehicle propulsion and to internal friction, would be returned to the desired rate). If used in a non-hybrid application, of course, EMBs would, like ECBs, have to be recharged using the standard electric power grid. They thus share the ECB property of being "emission moving" rather than "zero emission" propulsion systems.

EMBs are a relatively new technology, still under development. Current costs are high, and materials issues remain to be resolved. Further research and development may well make EMBs an important part of the energy systems of automobiles.

9.5.3   Hybrid-Powered Vehicles

Some of the performance disadvantages of electric vehicles can be overcome in principle by providing hybrid vehicles in which battery power is used for cruising and an alternative power source provides added energy for acceleration. Several alternative power sources are being evaluated: fuel cells, compressed natural gas engines, and small gasoline and diesel engines.

The one that appears most promising, at least for the near term, uses an internal combustion engine. This engine has, of course, the advantages of decades of design refinement and good performance under load, and is sized for average rather than peak power requirements so that its efficiency is optimized. The electrical system recovers braking energy rather than dissipating it. A bonus is that any hybrid vehicle, but especially internal combustion hybrids, can add range to the vehicle, thus ameliorating one of the principal disadvantages of the electric car. An important, if subtle benefit of internal combustion hybrids is that they begin to provide a graceful transition away from total dependence on the petroleum infrastructure.

Hybrid vehicles are not zero-emission vehicles, but operate with substantially decreased emissions. From a mechanical standpoint, however, they are significantly more complicated than either electric or gasoline engines by themselves. It remains to be seen whether developments are sufficiently promising that the advantages will appear to outweigh the disadvantages.

9.5.4   Fuel Cell-Powered Vehicles

A fuel cell is an electrochemical energy provider that has some similarities both to electrochemical batteries and to gasoline engines. Fuel cells use hydrogen or methanol stored in on-board tanks rather than chemicals within them (as do ECBs), but the energy-producing reaction is electrochemical. A conceptual sketch of a hydrogen fuel cell is shown in Fig. 9.6. At the anode, the stored hydrogen gas is ionized in a reaction that can be written as

$$2\,H_2 \rightarrow 4\,H^+ + 4\,e^-$$

The hydrogen ions then permeate a membrane and react with oxygen at the cathode by

$$O_2 + 4\,H^+ + 4\,e^- \rightarrow 2\,H_2O$$

The resulting charge difference generates an electric current to provide power to the vehicle. The emission is restricted to water vapor. In the methanol analogue, the emissions are still relatively benign, because the low operating temperature of the fuel cell does not result in the formation of highly reactive smog-forming fragments.

Compared to ECBs, fuel cells have much higher energy density, and refueling can be much faster. Although their use involves the removal of fuel from tanks, its transformation, and the emission of combustion products, those products, largely water, are benign. Considerable energy is required to generate the hydrogen, however, and this energy is provided remotely from the vehicle by the electric power grid. Hence, the "emission moving" label appropriate to batteries applies also to fuel cells, together with some attributable emissions of acid gases such as sulfur dioxide.

The use of fuel cells burning hydrogen to power vehicles is attractive from the standpoint that the only product of the combustion is water vapor. The hydrogen can be produced from many sources. The most straightforward is a process from natural gas, but this requires the use of a fossil fuel resource in increasingly short supply. Generation from water by electrolysis, with solar or hydropower providing the energy, is an attractive option providing the renewable power is available, but the process is expensive. In addition, implementation of the fuel cell option would probably require vehicles to carry with them a pressurized tank of flammable hydrogen, thus posing a possible safety problem as well as a

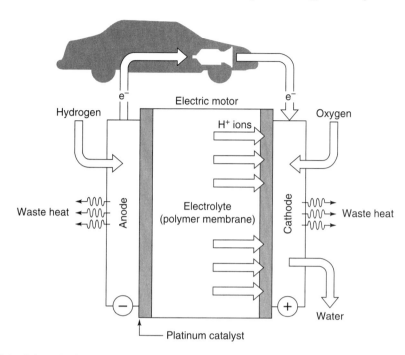

Figure 9.6   Schematic diagram of a fuel cell. (Adapted with permission from P.M. Eisenberger, *Basic Research Needs for Vehicles of the Future*, Princeton, NJ: Princeton Materials Institute, 1995.)

problem in vehicle design. (This problem is not insurmountable, however, as the fuel management systems of current automobiles demonstrate.) Extensive research is under way to explore alternative and safer methods of storage, as in ceramic sponges that prevent flammability but permit the hydrogen to be readily recovered and used; nonetheless, the volume inefficiency of the storage will almost certainly be a continuing liability.

9.6 ENERGY CONSUMPTION IN AUTOMOTIVE INFRASTRUCTURE MANUFACTURE

Substantial amounts of energy are required to manufacture the automotive infrastructure—roadways, bridges, signage, etc. The amount of energy invested in producing one gram of various construction materials ready for use (their "embedded energy") is given in Table 9.3. Bitumen is manufactured relatively efficiently from crude oil, and has a modest embedded energy. That for cement, which requires substantial heating in its manufacture, is about ten times higher. Steel, which is made by the recovery of iron ore, high heating to molten metal, alloying, and processing, is three times higher still. The embedded energy of aggregate (crushed rock), although high, is by far the lowest of any material in the table.

Table 9.3   Embedded Energy in Roadway Construction Materials*

Material	Embedded Energy (J/g)
Bitumen	630
Cement	6700
Aggregate	74
Reinforcing steel	2.3×10^4
Steel beams	1.8×10^4

*The data are from Stammer, R.E., Jr., and F. Stodolsky, *Assessment of the Energy Impacts of Improving Highway-Infrastructure Materials*, Rpt. ANL/ESD/TM-115, Argonne, IL: Argonne National Laboratory, 54 pp., 1995.

The embedded energy in the world's roadway infrastructure can be calculated by combining the information on Table 9.3 with that on roadway type and distance. The results are given in Table 9.4. Some estimation is certainly involved, so the numbers should not be regarded as highly accurate, but it appears that the minimum amount of embedded energy is 190 EJ and the true number may be 20–30 percent higher when all aspects of the infrastructure are accounted for. (For perspective, the total world energy use for all purposes in 1990 was 390 EJ, of which about 20% was consumed in transportation fuel use. The embedded energy in the world's roadway infrastructure is thus roughly half a year's world total energy consumption.)

9.7 SUMMARY

Alternative energy supplies for automobiles are all rather problematic as this is written. Technologies are not fully developed, and even the technologically adequate seem likely to be implemented only at significant additional cost to the consumer, together with the possibility of degraded performance, at least in the short term. The most promising concept appears to be the hybrid vehicle, probably using lead-acid batteries and a small internal combustion engine, at least for the next decade. Extreme reliability and elimination of cost differentials will be required if hybrid vehicles are to capture a significant share of the world vehicle market. Over the longer term, fuel cells seem likely to play an important role.

Unlike the automobile, for which energy-saving alternatives are under development, there appears to be little prospect for decreasing energy use as it applies to roadways, parking lots, parking garages, and other automotive-related structures. The energy embedded in infrastructure thus seems likely to continue to be very large, especially in the developing countries where infrastructure improvement has a high priority.

Table 9.4   Embedded Energy in the World's Roadways

Rural roadways:

Energy per lane distance	8.4 TJ/km[*]
Energy per 12 m bridge	1.6 TJ[*]
World road distance	9.5×10^6 km[‡]
Embedded roadway energy	170 EJ[+]
Embedded bridge energy	0.73 EJ[@]
Total embedded energy	170 EJ[&]

Urban roadways:

Energy per lane distance	10.0 TJ/km[*]
Energy per 12 m bridge	1.6 TJ[*]
World road distance	1.1×10^6 km[‡]
Embedded roadway energy	22 EJ[+]
Embedded bridge energy	0.80 EJ[@]
Total embedded energy	23 EJ[&]

World roadways:

Embedded roadway energy	190 EJ
Embedded bridge energy	1.5 EJ
Total embedded energy	190 EJ

[*] The data are from Stammer, R.E., Jr., and F. Stodolsky, *Assessment of the Energy Impacts of Improving Highway-Infrastructure Materials*, Rpt. ANL/ESD/TM-115, Argonne, IL: Argonne National Laboratory, 54 pp., 1995.

[‡] The data are from International Road Federation, *World Road Statistics, 1985–1989*, Geneva, 1990.

[+] Assumes 2.1 lanes per roadway.

[@] Assumes 1 bridge/2 km (urban), 1 bridge/20 km (rural).

[&] Not included is embedded energy from lighting, signage, guard rails, fences, traffic signals, etc.

SUGGESTED READING

Bleviss, D.L. and Walzer, P., Energy for motor vehicles, *Scientific American, 263* (3), 102–109, 1990.

Linden, H.R., Energy and industrial ecology, in *The Greening of Industrial Ecosystems*, B.R. Allenby and D.J. Richards, eds., Washington, D.C.: National Academy Press, 1994.

MacKenzie, J.J., *The Keys to the Car: Electric and Hydrogen Vehicles for the 21st Century*, World Resources Inst., Washington, D.C., 128 pp., 1994.

National Research Council, *Automotive Fuel Economy: How Far Should We Go?*, National Academy Press, Washington, D.C., 1992.

O'Connor, L., Electric vehicles move closer to market, *Mechanical Engineering, 117* (3), 82–87, 1995.

Post, R.F., A new look at an old idea: The electromechanical battery, *Science and Technology Review*, pp. 13–19, April, 1996.

Smith, J.R., Getting along without gasoline—The move to hydrogen fuel, *Science and Technology Review*, pp. 28–31, March, 1996.

Wouk, V., Hybrids: Then and now, *IEEE Spectrum 32*(7), 16–21, 1995.

EXERCISES

9.1   An automotive part, made entirely from plastic, is molded in the factory from polymer resin. The sequence of manufacturing steps is: injection molding, trimming and smoothing, inspection, and shipment. In each of the first three steps a portion of the material is consigned to waste disposal. Steps 2 and 3 are, however, able to recycle some of the wastes back to Step 1 for reprocessing. Draw process flow sheets for (1) polymer/plastic and (2) energy. Label the flows and write an equation for the overall efficiency of use of the polymer material.

9.2   Give examples of geographical regions for which electric vehicles are particularly well suited. Why are some regions better choices than others? Are competing modes of transportation better options than either gasoline or electric vehicles? Why?

9.3   A four-lane connector roadway is proposed to link two highways in rural areas that are 15 km apart. Five 24 m bridges will be required. What will be the embedded energy in the connector roadway? Repeat the calculation for the case where the roadway traverses an urban area.

9.4   (a) Compare the cost in Germany for a unit of energy produced by gasoline with that of a unit of energy produced by electricity. The energy released by gasoline is 43 MJ per liter. Assume current prices to be 0.03 Deutschmarks per MJ for electricity and 1.6 Deutschmarks per liter for gasoline. Do the same for the U.S., where electricity is 5c per kWh and gasoline is 150c per gallon. (b) Gasoline engines in automobiles are about 12% efficient, and electric cars are about 55% efficient. Which is the more economical means of transport in Germany so far as fuel is concerned?

9.5   The energy produced from petroleum in the U.S. in 1990 was about 35 EJ, and 46% of the petroleum was supplied from foreign sources. 67% of the petroleum was consumed in transportation. Overall automotive fuel economy in the U.S. was about 8.9 km/l, as opposed to 11.4 km/l in Japan. If the U.S. fleet had averaged 11.4 km/l, how much less petroleum would have had to have been imported? Repeat the calculation for a fleet average of 40 km/l, the fuel economy of the best prototype vehicles.

CHAPTER 10

Environmental Interactions During Product Use

"The principal environmental impacts of the automobile occur during its use by the customer, and any assessment that is restricted to manufacturing misses the most significant life stage."
— Inge Horkeby, *Volvo Car Corporation*

10.1 THE EMPLOYMENT OF MATERIALS

Many industrial products require materials while they are being used as well as while they are being manufactured, and such products include automobiles and their supporting infrastructure. The material most voraciously consumed during product use is, of course, gasoline. In Chapter 9 we noted that the world's vehicle fleet currently uses about half of all global petroleum production. Many other materials are involved in the customer use of automobiles: lubricants, coolants, materials in replacement parts, tires, batteries, and so on.

The infrastructure in-use stage also involves the employment of materials. Road patching, renovation, and replacement is a continuing process, as is the case with bridges. There are a host of aperiodic activities involving materials: the replacement of signage, updating lighting and signalling, installation of new roadway drainage systems, and so on. Finally, there are the true consumable uses of materials in automotive infrastructures: lines on roadways are regularly repainted and, under conditions of ice and snow, sand and salt are spread on roadways to increase traction.

The use of consumable materials in the maintenance of automobiles and infrastructure implies the continuing flow of those materials. Road salt, for example, is dug from its reservoirs, transported to where it is used, spread across the roadways, and eventually washed to the nearest stream or water treatment plant, probably causing ecological damage in the process. Other consumables are mostly recycled (as with used engine oil, for example), but the recycling process invariably involves energy consumption, at least.

The scale of environmental interactions during product use is more important than it might at first appear, because automobiles are very long-lived compared with many other industrial products and because the flows of materials and energy connected with them are so large. When an automobile is built its environmental interactions occur over a week or two, and most of those impacts are under fairly intensive government control and oversight. Once in service, that same automobile may last for ten or fifteen years, interacting with the environment in a number of ways (as we shall see) every day it is in operation. Consequently, it is the in-use stage where the greatest impacts of the automobile are to be felt.

Roadways are even longer-lived than automobiles. Should a roadway be designed and built so as to require frequent patching, or more than the average amount of road salt to keep it navigable, or supplementary drainage to prevent corrosion of the reinforcing steel, the environmental effects may be felt for decades. Unlike the designer of the portable radio or the umbrella, whose mistakes will not matter a year or two after purchase, the automobile or roadway designer's blunders live on year after year. Conversely, so do her or his design triumphs.

10.2 SOLID RESIDUES FROM AUTOMOBILE USE

Old tires are a good object lesson in disposal and reuse. Although tire consumption per vehicle has decreased by a factor of two or three in the past 20 years because modern tires last much longer, the numbers of discarded tires are still daunting—the United States alone throws away 250 million every year. For decades these tires were dumped in landfills and other less suitable places, but limited landfill capacity, research into a variety of uses for discarded tires, and a feeling that there must be better alternatives is gradually changing that approach.

Retreading tires is useful in lengthening the service life and is probably environmentally preferable to discarding the tires after a single use, though we know of no study of the topic. Nonetheless, retreading of tires merely delays the inevitable. Upon eventual discarding, a fraction of today's old tires are sent to modern facilities that shred and separate them into three flow streams: small tire chunks, steel shards, and crumbs. The steel is readily recyclable. The crumbs are burned for energy (each tire contains more than eight liters of recoverable petroleum). The chunks see a variety of uses—for running tracks, rubber boots, and rubberized asphalt, to name a few. As a possible alternative to the shredding operation, the British company BOC Ltd. has developed a method to recycle scrap tires by first freezing them with liquid nitrogen and then grinding them efficiently while in the frozen state. Methods to turn the carbonaceous fraction of tires back into petroleum are being explored as well.

Although adequate technology is becoming available for tire recycling, appropriate economics may not yet be in place. Tire disposal costs have never been internalized; unlike aluminum beverage cans, tire users do not currently pay a recyling deposit upon initial purchase. The result is that any recycling that occurs happens only if it is profitable, and in many cases it is not. Part of the reason for this situation is that legislation often prohibits the use of tires as feedstock for incinerators (though the petroleum from which synthetic rubber is made is an approved fuel) or makes it difficult to transport old tires into a city or across a state line to where a recycling facility is located. It is obvious that the recycling industry, economists, and politicians all have roles to play if the issue is to be properly addressed.

The tire design engineer also has a role to play: designing tires for the environment. Today's recyclers are dealing with tires designed with no consideration for their eventual disposal. Perhaps a tire's composition can be changed to make it burn more efficiently while releasing few or no toxics. Perhaps a tire can be made so as to be more quickly and easily separated into its components. Perhaps a tire can be reformulated so as to be more readily transformed into a new product. In the case of old tires, the recycling engineers have made substantial progress and the economists and politicians are beginning to think about the situation. The design engineer has yet to start.

Batteries are a recycling success story, with more than 90% of the lead in the batteries being recovered and reused. This occurs because the lead is reasonably valuable, because battery designs permit efficient reprocessing of the lead, and because the battery infrastructure is highly developed. Nonetheless, some 10% of the lead in motor vehicle batteries is thought to be lost either to non-recovered batteries or to loss during reprocessing. This lost lead may have significant environmental impacts, and its magnitude and effects need to be studied in detail.

Automobile parts form another materials stream of significant proportions, and again it is a stream that is relatively well handled from an industrial ecology standpoint. The infrastructure for reuse of automobile parts is one of the best in the industrial system, and many parts can be readily rebuilt and restored to service. No estimates of quantities or efficiencies are available for parts recycling, and certainly parts designs could be made easier to deal with at end-of-life, but the materials in automobile parts are certainly among the better used of the streams connected with motor vehicles.

Rough estimates of the total amounts of solid residues attributable to the in-use stage of the automobile life cycle are given in Table 10.1. For rubber and lead, the automotive fraction is sizable. In the case of defective parts no figures are available, but they must surely be only a small fraction of the steel, plastic, and aluminum waste of society as a whole.

10.3 LIQUID RESIDUES FROM AUTOMOBILE USE

The liquid residues from automotive in-service use are not enormous in magnitude, but have toxic concerns. They include the variety of lubricating oils and fluids contained in engines, transmissions, and power steering reservoirs, as well as radiator coolants. Motor oil is the fluid in largest use, several liters being cycled through the average vehicle each year. It is estimated that 80% or more of the used oil is collected, filtered, reprocessed, and reused, with the remainder being lost to dissipation and dumping. Antifreeze agents, most commonly ethylene glycol, are of particular interest in that consumers often change or supple-

Table 10.1   Residues from the Product Use Stage of the World Vehicle Fleet

Material	Annual flow per vehicle	Annual flow per global fleet	Ref.	Annual flow per anthro. sources	Ref.	Auto percent of global flow
Solid Residues						
Rubber (Tires)	7 kg	4.5 Tg	GA97	14.5 Tg	R93	31
Lead (Batteries)	5 kg	2.6 Tg	GA97	5.8 Tg	TS94	45
Parts	?	?				
Liquid Residues						
Oil	10 l	8.3 Tg	GA97			
Antifreeze	2 l	0.9 Tg	GA97	5.4 Tg	GA97	17
Gaseous Residues						
CO_2		1.0 Pg C	HO94	5.8 Pg C	A96	16
NO_x		8.6 Tg N	HO94	21 Tg N	B96	41
CH_4		0.4 Tg C	HO94	535 Tg C	IPCC95	0.1
NMHC		22 TgC	HO94	88 Tg C	HO94	25
CO		51 Tg C	HO94	200 Tg C	HO94	25
N_2O		15 Gg N	HO94	15 Tg N	IPCC95	0.1
SO_2		3–6 Tg S	NRC93	65 Tg S	B96	5
Black carbon		0.6 Tg C	P93	12 Tg C	P93	5
Lead		120 Gg	P95	180Gg	P95	67

A96: R.J. Andres, G. Marland, I. Fung, and E. Matthews, A one degree by one degree distribution of carbon dioxide emissions from fossil fuel consumption and cement manufacture, *Global Biogeochemical Cycles, 10*, 419–429, 1996.

B96: C.M. Benkovitz, M.T. Scholtz, J. Pacyna, L. Tarrason, J. Dignon, E.C. voldner, P.A. Spiro, J.A. Logan, and T.E. Graedel, Global gridded inventories of anthropogenic emissions of sulfur and nitrogen, *Journal of Geophysical Research, 101*, 29, 239–29, 253, 1996.

GA97: This work

HO94: J. Hahn and B. Oudart, *Contributions of the Automotive Industry to the Global Emissions of Greenhouse Gases: Present State of Knowledge*, Garmisch-Partenkirchen: Fraunhofer Institut für Atmospharische Umweltforschung, 1994.

IPCC95: Intergovernmental Panel on Climate Change, *Climate Change 1994*, Cambridge, U.K.: Cambridge University Press, 1995.

NRC93: Committee on Haze, *Protecting Visibility in National Parks and Wilderness Areas*, Washington, D.C.: National Academy Press, 1993.

P93: J.E. Penner, H.E. Eddleman, and T. Novakov, Towards the development of a global inventory of black carbon emissions, *Atmospheric Environment, 27A*, 1277–1295, 1993.

P95: J.M. Pacyna, M.T. Scholtz, and Y.-F. Li, Global budget of trace metal sources, *Environmental Review, 3*, 145–159, 1995.

R93: D.G. Rogich, United States and global material use patterns, Paper presented at ASM International Conference on Materials and Global Environment, Washington, D.C., Sept. 13, 1993.

TS94: V. Thomas and T. Spiro, Emissions and exposure to metals: Cadmium and lead, in *Industrial Ecology and Global Change*, R. Socolow, C. Andrews, F. Berkhout, and V. Thomas, Eds., Cambridge, U.K.: Cambridge University Press, 1994.

ment their own coolant reservoirs, and the spent fluid is more likely to see unwise disposal than if it is recovered by a professional mechanic. Coolant fluids tend to be difficult to recycle because of the metal particles they customarily contain, but active efforts are under-way.

Rough estimates of the total amounts of liquid residues attributable to automobiles are given in Table 10.1. While a large quantity of lubricating oil is used by motor vehicles, much of it is recycled and much oil is used elsewhere in a wide variety of uses, so the motor vehicle fraction must be small. For ethylene glycol (antifreeze), the fraction of global use for all purposes is between 15 and 20 percent. For completeness, we mention without esti-mation the water and detergents used in motor vehicle washing facilities. In many cases this resource is recycled; sometimes it is merely dispersed into the environment.

10.4 GASEOUS RESIDUES FROM AUTOMOBILE USE

Automobiles burn petroleum, a mixture of hydrocarbons that may be represented by the generic formula C_xH_y. Complete combustion of pure hydrocarbons produces only carbon dioxide and water, by combination with oxygen in air:

$$C_xH_y + O_2 \xrightarrow{\Delta} x\, CO_2 + \frac{y}{2}\, H_2O$$

where the Δ over the arrow indicates that the reaction occures under high temperature con-ditions. A simultaneous reaction that takes place at the high temperature of combusion is the combination of oxygen and nitrogen from the air to form nitric oxide:

$$N_2 + O_2 \rightarrow 2\, NO$$

Because of imperfect fuel mixing, temperature variations, and other factors, the combustion reaction does not generally go to completion, but produces varying amounts of carbon monoxide (CO), methane (CH_4), unburned or partially oxidized hydrocarbons (des-ignated non-methane hydrocarbons, or NMHC), and black carbon particles, along with the CO_2, H_2O, and NO. Table 10.1 presents the estimated global emissions from automobiles, together with their fractional contributions to global budgets.

Computer control of combustion and the fitting of exhaust catalysts has markedly reduced emissions of these secondary combustion products. Combustion efficiencies differ greatly with level of control, however, as well as with the age of the vehicle and its level of maintenance. And, even if combustion were perfect CO_2 and H_2O are still generated.

As shown in Table 10.2, tetraethyl lead is added to gasoline in many countries as an octane enhancer. This use is about 2% of the global lead use for all purposes, and is com-pletely dissipative, resulting in the transfer of lead to the environment in the form of lead-containing aerosol particles that settle rapidly to the ground, becoming part of the dusty urban environment.

Sulfur is also present in crude oil as a natural constituent, though many refineries treat the petroleum so as to greatly reduce the sulfur content. At combustion, the sulfur is con-verted into a variety of oxygenated products; by far the most abundant of these is sulfur

Table 10.2   Use of Leaded Motor Vehicle Gasoline, 1990

Country	Gasoline Consumed (Gl/yr)	Percent Leaded	Lead Content (g/l)	Lead Emitted (Gg/yr)
Eastern Europe, Asia				
FSU	110	95	0.2	20
Poland	3.7	100	0.4	1.5
Hungary	2.2	100	0.4	0.9
Central, South Amer.				
Venezuela	9.9	100	0.85	8.4
Colombia	6.4	100	0.8	5.1
Argentina	5.3	100	0.4	2.1
Western Europe				
France	27	85	0.25	5.8
Italy	19	95	0.3	5.5
Spain	11	99	0.4	4.4
United Kingdom	37	30	0.15	1.7
Sweden	5.7	50	0.15	0.4
North America				
Mexico	26	90	0.2	4.6
United States	420	2	0.026	0.2
Far East, Oceania				
China	27	46	0.01–0.48	0.1–5.9
India	4.8	100	0.56	2.7
Thailand	3.7	100	0.4	1.5
Africa				
South Africa	6.4	100	0.3	1.9
Egypt	3.0	100	0.4	1.2

Sources of data for this table: gasoline consumption from U.S. Dept. of Energy, 1991; former Soviet Union gasoline lead content from Russian Standard, 1987; other lead contents from Octel, 1992.

dioxide (SO_2). SO_2 emissions from motor vehicles are, however, only a minor component of the global sulfur budget.

A final gaseous emittant from automobiles is the refrigerant used for air conditioning. Leaks in automotive air conditioning systems are not uncommon, and escaped refrigerant is dispersed into the air. For decades, the automotive refrigerant used was CFC-12. Since the early 1990s, HFC-134a has been used. Data for the quantities used are not available.

Each of the emittants from automobile exhaust has a substantial environmental impact. CO_2 is a major contributor in global climate change, as is methane (and, to a lesser degree, CO and SO_2). NO_x (NO and its oxidized variant NO_2) and NMHC are involved in the generation of photochemical smog. SO_2 and NO_x are the precursors of acid rain. CO and lead are toxic to humans and other organisms, and black carbon and NO_2 decrease visibility. These impacts, and the contribution of automotive emissions to them, are described later in this chapter.

The gaseous residues from motor vehicles are compared with global budgets of the emittants in Table 10.1. In several cases, especially NO_x, CO, non-methane hydrocarbons, and lead, the fraction chargeable to vehicles is a substantial portion of the global totals.

10.5   ENVIRONMENTAL IMPACTS FROM NON-DESIGN USE

10.5.1   Malfunctioning Vehicles

Most well-maintained vehicles meet or better the levels of exhaust emissions control for which they were designed. In a few cases, however, individual vehicles are gross polluters, and these vehicles can represent any model year including the newest. In a classic study demonstrating this property, researchers from the University of Denver monitored the emissions from more than 65,000 vehicles. Their findings, shown in Fig. 10.1, indicated that 5–10% of the vehicles accounted for more than 50% of the carbon monoxide and NMHC emissions from the entire vehicle sample. Although the oldest gross polluting vehicles

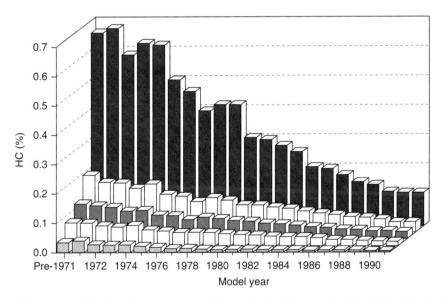

Figure 10.1 Average NMHC exhaust emissions (relative scale) of vehicles monitored in California by the University of Denver. The emissions for vehicles from each model year were sorted and divided into five equal groups. The average emissions rate for each of the five groups for each model year is plotted from front to back, lowest to highest. (Reproduced by permission from Beaton, S. P., G. A. Bishop, Y. Zhang, L. L. Ashbaugh, D. R. Lawson, and D. H. Stedman, On-road vehicle emissions: Regulations, costs, and benefits, *Science, 268*, 991–993, 1995.)

were the worst, recently manufactured gross polluting vehicles had emission rates higher than those of half of all vehicles twenty years older.

Examination of gross polluting vehicles showed in almost every case that the vehicles were poorly maintained or had been tampered with (probably illegally). Thus, improvements in auto-related air quality are most likely, and can be achieved most efficiently, if government programs attempt to identify and repair or scrap the gross polluters rather than to gradually tighten regulations across the entire vehicle population.

10.5.2   Non-Optimum Vehicle Operation

Emissions are related not only to total distance driven, but also to the way a vehicle is operated. A major factor is the average distance per trip, since emissions at startup are as much as one-third those of an entire trip. A series of short trips thus has much more environmental impact than a single longer trip of equivalent distance.

A second operational factor is the average vehicle speed. Emissions are highest in slow, congested traffic conditions, lower at intermediate speeds, and higher again at high speeds because of increasing wind resistance (but not as high as at the slowest speeds). Thus, the congestion so common in today's cities strongly affects local, regional, and global air quality. In many major cities, including London, New York, Paris, and Tokyo, average speeds during rush hour periods are often less than 10 km/hr, and the vehicle emissions are therefore high (unavoidably, so far as the individual driver is concerned). This phenomenon reinforces the discussion in Chapter 4 of the need to view the system as a whole: optimizing the efficiency of each individual vehicle does little for overall emissions unless the performance of the transport system as a whole is also improved.

A third factor producing increased emissions is rapid acceleration. This results because engines are designed to use higher fuel-to-air ratios under acceleration to provide satisfactory driving performance. Among the consequences of this strategy is insufficient oxygen for good catalytic converter performance and thus a greater throughput of unprocessed exhaust products.

10.6   IN-USE INFRASTRUCTURE IMPACTS

10.6.1   Airborne Particle Generation

As mentioned above for automobile exhaust emissions, airborne particles can produce deleterious health effects, as well as decreased visibility. These impacts are important in connection with the roadway infrastructure, since that infrastructure is the largest source of particle emissions. As seen in Fig. 10.2, unpaved and paved roads are responsible for nearly half of all emissions. If a quarter of the construction emissions are assigned to transportation, the total exceeds 50%.

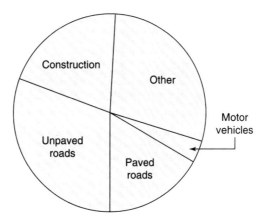

Figure 10.2   Selected sources of particulate matter with diameters less than 10 µm, United States, 1992. The total estimated US particle emissions from all sources are about 47 Tg/yr. (Data from S. Nizich, Ed., *National Air Pollutant Emission Trends, 1900–1992*, Rpt. EPA-454/R-93–032, Research Triangle Park, NC: Environmental Protection Agency, 1993.)

10.6.2   Road and Highway Maintenance

Once an automotive infrastructure network is in place, it cannot, of course, be forgotten. Snow must be plowed from the roads, sand and salt are spread on ice, urban streets, at least, must be periodically cleaned, new center lines must be painted. Weather, corrosion, road wear, and accidents require a continuing maintenance effort. All of this activity involves the movement and use of a continuing spectrum of materials.

Given the existence of a roadway network, most of the maintenance activities mentioned above are both reasonable and, from an environmental standpoint, relatively benign. An exception is a notable new feature of infrastructure maintenance in the last quarter century: the spreading of materials on roadways during winter to improve traction in ice and snow. First begun with sand, the preferred material is now salts such as sodium and calcium chloride that depress the melting point of water and maintain liquid conditions several degrees below 0°C. Concerns for safety and convenience have sharply increased the rate of use of deicing salts in recent years, as shown in Fig. 10.3 for the United States.

There is a substantial cost connected with supplying and applying the deicing salts, of course, but two potentially greater costs are harder to quantify. One is the much enhanced corrosion of vehicles, bridges, reinforcing steel, and other materials; as a result, automobile manufacturers have sharply increased their use of zinc plating of body panels in recent years, producing a decrease of the global zinc supplies as well as a decrease in the horizontal recycling of steel. The second is cost related to the impacts of sand and salts on the ecosystems near the roadways. Forests, grasslands, wetlands, and parks are all degraded by sand and salts washed into them from the highways.

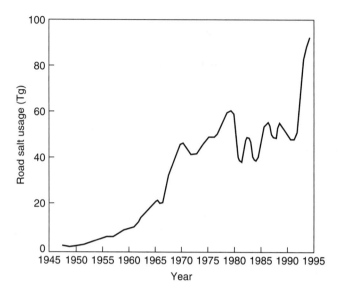

Figure 10.3   The use of road salt in the United States, 1945–1995. (Adapted from a diagram devised by R. Baboian, from data provided by the Salt Institute, Washington, D.C.)

   The environmental impacts of most infrastructure maintenance activities are relatively local, but that of the manufacture of cement is global as well. Cement is the manufactured material with the largest global consumption rate, about 1.25 Pg in 1990. A substantial fraction of this material is used to build and maintain the transportation infrastructure. The production of cement is energy-intensive and results in the emission of about 3% of the global anthropogenic carbon dioxide, so cement production plays a modest role in global climate forcing. Research to decrease this impact seems at least as logical as trying to design a motor vehicle to get 10% more distance per unit amount of fuel, an activity that has been vigorously persued over the past two decades.

   Finally, there are a number of lesser impacts of infrastructure maintenance that deserve mention. The yellow lines so common on roadways are lead chromate. This material appears relatively benign in use, but its manufacture requires the processing of lead and chromium, both of which are toxic, and its degradation on roads over time results in the dispersion of these toxic metals into the environment. Other in-use impacts include the consumption of electrical power for lighting, signaling, and the ventilation of traffic tunnels. As with other products in continual use, even an increase in efficiency of a few percent can have significant long-term impacts on total power consumption.

10.6.3   Storage of Motor Fuels

Motor fuels are stored in (mostly underground) tanks at distribution centers and petrol stations, and many of those tanks are old and corroded. As a result, leakage from underground storage tanks is very common. Overall world storage tank figures are not available, but a recent study by the American Petroleum Institute and the U.S. Environmental Protection Agency derived a total number of active underground storage tanks for gasoline, diesel fuel, and heating oil in the U.S. of about 1.1 million. Of these, nearly 30 percent have had confirmed releases. This is a continuing problem for local and regional water supplies, as a number of potentially carcinogenic constituents are contained in gasoline, diesel fuel, and other petroleum derivatives. New tank materials and increased vigilance in monitoring tank levels are reducing leakage frequencies and leakage rates, but tank leakage will surely remain a significant environmental impact for years to come.

10.7 GLOBAL CLIMATE CHANGE AND THE AUTOMOBILE

The basic source of energy for most terrestrial processes, and certainly for climate, is the absorption of solar radiation. If the climate system is to be in equilibrium, the solar radiation that is absorbed by Earth must be balanced by outgoing thermal radiation from Earth to space. The global radiation budget is shown in Figure 10.4.

An average of slightly less than 30% of the incoming solar radiation received by Earth is returned into space: by reflection (backscattering) from clouds (about 19%), backscattering by air molecules and particles in the air (together about 6%), and reflection at Earth's surface (about 3%). Almost 25% of the solar radiation is absorbed within the atmosphere, mostly by ozone in the stratosphere (about 3%), and by clouds (5%) and water vapor (17%) in the troposphere. The remaining 47% of the incoming solar radiation is absorbed at Earth's surface.

Of the solar energy absorbed at Earth's surface, a little more than half goes into *latent heat*, that is, heat absorbed by water as a consequence of its transformation from liquid form to vapor form at Earth's surface and released again into the atmosphere when water droplets condense in clouds. Other significant amounts of surface heat energy are transferred to the atmosphere by convection and turbulence (about 10%) and by the absorption of infrared radiation by the greenhouse gases. Of the 47% of the initial solar energy absorbed at Earth's surface, only 18% is lost by radiation, because the remainder is captured in the atmosphere. This atmosphere-surface cycling is the greenhouse effect; it causes Earth's surface to be about 33 °C warmer than would otherwise be the case. An increase in the concentration of a greenhouse gas initially decreases the flux of long-wave terrestrial radiation to space as more of the radiation is trapped in the troposphere. The effect will be a temperature rise at Earth's surface, the magnitude depending in part on related process such as changes in water evaporation rates or cloudiness.

Much of the current concern over the impact of human activities on climate arises from the ability of human development of Earth to alter the amount of absorbed or emitted radiation, or to change the hydrologic cycle of the planet. In assessing these impacts, one must place them in perspective with natural variations in climate and driving forces. From the point of view of Earth's radiation budget, the most important gas is water vapor, whose atmospheric concentrations cannot be directly influenced by human activity to a large degree. Next in importance is carbon dioxide, whose effect is calculated to be large over

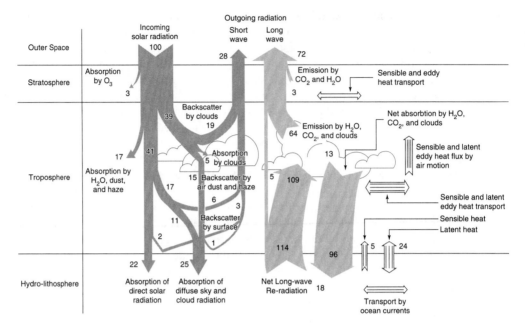

Figure 10.4   The annual mean global energy balance for the Earth-atmosphere system. *Sensible heat* is that transferred to the atmosphere from the heated surface by turbulent eddies; *latent heat* is that supplied to the atmosphere upon condensation of water vapor. The figures are percentages of the energy of the incoming solar radiation. (Courtesy of S. H. Schneider.)

the next century. The combustion of gasoline and other fossil fuels is the most important factor in the steady increase in the atmospheric concentration of CO_2.

Carbon dioxide from motor vehicles might be termed the "invisible emittant". It is odorless, colorless, and spewed from the tailpipe far enough behind the driver so as to be easily ignored. The quantities of carbon dioxide emitted, however, are quite large. Burning one liter of gasoline produces about 2 kg of CO_2, and a vehicle with total lifetime travel of 160,000 km emits about 32 Mg of CO_2 during its period of use. All told, the world's vehicles emit each year about 1 Pg C as CO_2, some 16% of all anthropogenic CO_2 emissions.

The automotive contributions to the emissions of other greenhouse gases are not nearly so important as that of CO_2. After CO_2 (and water), the next most significant greenhouse gases are the CFCs. Where a few years ago automobile air conditioning with CFC-12 was responsible for about 15% of all CFC production, the recent substitution of HFC-134a as a refrigerant has decreased automobile CFCs as a greenhouse gas issue. CH_4 and N_2O, also potent greenhouse gases, are emitted at very low rates by automobiles. Automotive emissions of CO, while of concern from a toxicological standpoint, are not very significant as a contribution to global warming.

10.8 PHOTOCHEMICAL SMOG AND THE AUTOMOBILE

In the early part of this century, ground-based measurements and in situ balloon-based observations made it apparent that most of the atmosphere's ozone (O_3) is located in the stratosphere, the peak concentration occurring at altitudes between 15 and 30 km. For a long time, it was believed that tropospheric ozone originated in the stratosphere and that most of it was destroyed by contact with Earth's surface. Ozone was known to be produced by the photodissociation of O_2, a process that can only occur at wavelengths shorter than 240 nm. Because such short-wavelength radiation is present only in the stratosphere, no tropospheric ozone production is possible by this mechanism. In the mid-1940s, however, it became obvious that production of ozone was also taking place in the troposphere. After heavy injury to vegetable crops had occurred repeatedly in the Los Angeles area, it was shown that plant damage could be produced by ozone, which was known to be a prominent constituent of photochemical smog. The overall reaction mechanism was eventually identified by Arie Haagen-Smit of the California Institute of Technology as

$$NMHC + NO_2 + h\nu \rightarrow NO + \text{other products} \qquad (10.1)$$

where NMHC denotes various reactive nonmethane hydrocarbons (ethylene, butane, etc.), the catalyst is NO_x ($NO + NO_2$), and $h\nu$ indicates a quantum of solar radiation of wavelength less than about 410 nm. Ozone formation by this mechanism results from solar radiation dissociating the NO_2 formed in Reaction (10.1):

$$NO_2 + h\nu \ (\lambda \leq 410 \text{ nm}) \rightarrow NO + O \qquad (10.2)$$

and the recombination of O with molecular oxygen then produces ozone:

$$O + O_2 + M \rightarrow O_3 + M \qquad (10.3)$$

Because this pair of reactions is often followed by

$$O_3 + NO \rightarrow NO_2 + O_2 \qquad (10.4)$$

the reaction sequence can generate ozone in abundance only if some of the NO is oxidized to NO_2 without removing ozone. An efficient method for doing so in the urban atmosphere is

$$RO_2 \cdot + NO \rightarrow NO_2 + RO \cdot \qquad (10.5)$$

the $RO_2 \cdot$ radicals being derived from many hydrocarbons emitted into the air. The consequence of this chemistry is that ground-level ozone concentrations depend on both nonmethane hydrocarbon (NMHC) and oxides of nitrogen emissions. Calculations for ozone concentrations in a number of different urban areas have been performed with computer models of different dimensions and complexities. An approach that has seen wide use has been to make numerous calculations for a variety of emissions scenarios and to plot selected portions of the results for ready analysis. Figure 10.5, by Robert Bilger of the University of Sydney, Australia, illustrates such an effort. The diagram refers to conditions in the city and suburbs of Sydney and is used in the following way. Morning rush-hour con-

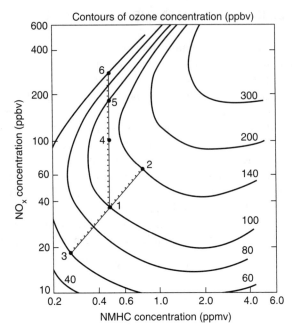

Figure 10.5   Isopleth diagram for maximum ambient daily ozone concentrations (in ppbv) derived for Sydney, Australia, given average morning rush-hour concentrations of NO_x and nonmethane hydrocarbons (NMHC). The numbers 1–6 on the figure are described in the text. (Reproduced with permission from R. W. Bilger, Optimum control strategy for photochemical oxidants, *Environmental Science and Technology, 12,* 937–940. Copyright 1978 by American Chemical Society.)

centrations of NO_x and NMHC are determined, often in the form of monthly, seasonal, or annual averages, to reflect the characteristic natural meteorological variations that influence concentrations. These NO_x and NMHC concentrations locate a point on Fig. 10.5. The curves on the figure are lines of constant concentration of the maximum ozone levels that the model predicts will be reached during a day in which the morning concentrations of the precursors are as specified.

It is instructive to study Fig. 10.5 for a situation where the NMHC and NO_x levels are changing. Consider the situation where the average concentrations are at point 1 on the figure. If emissions of both NO_x and NMHC increase and their concentrations move to point 2, the maximum ozone concentration will increase from about 100 ppbv to about 140 ppbv. (We say "about" here because the models are generally not considered accurate enough to give more than approximate results.) Conversely, if emissions of both precursors are reduced enough so that the concentrations move to point 3, the maximum ozone level will drop to about 60 ppbv. Now consider what happens if NMHC emissions are held constant while those of NO_x increase. If the concentration point goes from 1 to 4, the ozone maximum will increase. If NO_x emissions increase further, to point 5, the ozone maximum will retreat to approximately its former level. If NO_x emissions increase still further, to point 6, the ozone maximum will retreat still more, to below the level of point 1.

It is generally the case that engineering or economic considerations limit the type and amount of emission controls that may be used, so movement in certain directions on Fig. 10.5 is constrained or not possible. For example, NMHC emitted from vegetation can be important in generating ozone in certain locations and seasons. In addition, Fig. 10.5 is only an approximation of a complex chemical reaction set combined with a full spectrum of meteorological variables and local conditions.

There is little question that automotive emissions are the major cause of photochemical smog. On a global basis, automobiles emit some 40% of all NO_x and a quarter of all NMHC—the two principal smog generators. Increasing control of per-vehicle exhaust emissions have been counterbalanced by increasing numbers of vehicles on the road and by a greater distance driven annually per vehicle. Although the resulting ozone concentrations have diminished somewhat, they remain much higher than desirable.

10.9 STRATOSPHERIC OZONE AND THE AUTOMOBILE

A major catalytic cycle for ozone destruction in the stratosphere is that involving chlorine from chlorofluorocarbons (CFCs) and other chlorine-containing gases. The cycle was proposed in 1974 by Mario Molina and Sherwood Rowland, both then of the University of California at Irvine. Its basis is that CFCs are chemically very stable molecules that are not lost in the troposphere by chemical reactions or deposition. Rather, over a period of a few years these gases move from their release points at the surface up into the stratosphere. Above about 20–25 km the available solar radiation is energetic enough to destroy them, thereby releasing chlorine atoms, which interact with ozone as follows:

$$Cl \cdot + O_3 \rightarrow ClO \cdot + O_2 \tag{10.6}$$

$$O_3 + h\nu \ (\lambda \leq 1140 \ nm) \rightarrow O + O_2 \tag{10.7}$$

$$ClO \cdot + O \rightarrow Cl \cdot + O_2 \tag{10.8}$$

$$Net: 2O_3 + h\nu \rightarrow 3O_2$$

The key to this sequence of reactions is that the $Cl \cdot$ is regenerated to continue its ozone destruction. Until it is removed by reaction with some other stratospheric molecule, a single chlorine atom can destroy several thousand ozone molecules.

The most potent gases for ozone depletion are CFC-11 ($CFCl_3$) and CFC-12 (CF_2Cl_2). The latter was the refrigerant of choice for automobile air conditioners from the time of their introduction in the late 1950s until the early 1990s. Large quantities of CFC-12 were emitted to the atmosphere from the often leaky lines and seals in the air conditioners. With the advent of evidence that ozone depletion was serious, alternative refrigerants were explored. The CFC-12 substitute of choice, which began to be used in motor vehicles in 1993 and is now ubiquitous, is HFC-134a (CH_2FCF_3). Because this gas has hydrogen atoms in its makeup, it can be broken down in the lower atmosphere rather than reaching the stratosphere, and has none of the potent chlorine atoms in any case. The gas does have a modest greenhouse impact, but much less than that of the gas it replaces.

Automobile air conditioners are designed to operate at a pressure specific to the refrigerant used. As a result, a unit designed for CFC-12 cannot use HFC-134a without substantial modification. In addition, the lubricants used with CFC-12 are incompatible with HFC-134a, the combination leading to rapid compressor wear. Consequently, CFC-12 will be needed for at least a decade to service units that leak refrigerant to the atmosphere and need to be recharged. CFC-12, no longer manufactured, is being recovered and stockpiled for that purpose.

10.10 ACID RAIN AND THE AUTOMOBILE

The first major modern chemical study of precipitation was conducted in the 1960s by Christian Junge, a German scientist on an extended appointment at the Air Force Cambridge Research Laboratory in Massachusetts. Junge analyzed rain from a number of locations in North America, readily detecting the presence of chloride, sulfate, and nitrate ions in the water. He also measured the hydrogen ion (H^+), whose concentration can be indicated by specifying the solution's acidity, or pH value. The reader will recall from high school chemistry that the pH scale ranges from 0 to 14, low pH values indicating high acidity (high concentrations of H^+) and high pH values indicating high alkalinity (low concentrations of H^+). In pure water the pH is 7, and there are equal numbers of H^+ and OH^-, or hydroxide, ions. Water does not remain pure for long in the atmosphere, however, because the soluble gas carbon dioxide (CO_2) is present at a concentration of about 0.035%. It dissolves in water, combines with the water, and dissociates to produce atmospheric droplets of about pH 5.6.

Most rain has a more acid pH than 5.6, largely because natural and anthropogenic nitrogen and (especially) sulfur species increase the acidity. The most abundant sulfur-containing gas to interact with rain is sulfur dioxide (SO_2), which dissolves and then reacts in solution to form sulfuric acid (H_2SO_4).

The nitrogen-related acidity in acid rain comes not from the nitrogen dioxide (NO_2) emitted by automobiles, because NO_2 is not very water-soluble, but from nitric acid (HNO_3), made from NO_2 by reactions in the gas phase.

Because of sulfur and nitrogen oxides, most rain near urban areas has pH levels nearer 4.0 than 5.6 (Fig. 10.6). Cloud and fog droplets are almost always even more acidic than rain. In some fogs, in fact, the pH of the droplets has been found to be as low as 1.7—close to that of battery acid! It is no wonder that vegetation and other materials exposed to such fogs deteriorate rapidly.

The acidity of precipitation tends to be dominated by the hydrogen ions related to sulfate. On average, perhaps only about 30% of the acidity is related to nitrate and hence to NO_x. As Table 10.1 shows, the emissions of sulfur from motor vehicles are a small fraction of total sulfur emissions. The situation with NO_x is quite different, with automotive emissions comprising about 40% of the global total. Overall, then, the automobile contribution to acid rain is 10–15% of the total acidity. In cities where automotive emissions dominate, however, especially if the gasoline burned has not been desulfurized, the percentage can be much higher.

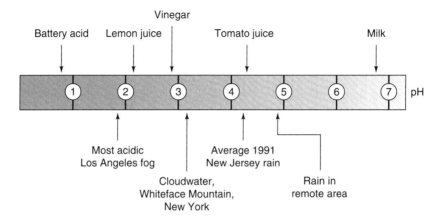

Figure 10.6   pH values in atmospheric water of various types, compared with pH values for several common liquids.

The discovery of acid rain was accompanied by the widespread assumption that the rain was causing substantial damage to soil, lakes, rivers, and structures. A decade or so of diligent research has confirmed some of these concerns and refuted others. The most severe and demonstrable damage appears to occur when highly acidic precipitation falls on surface waters within fragile ecosystems. The most sensitive of these systems are those of Scandinavia and northeastern North America, areas where the soil is thin and the organic matter that can absorb and neutralize the acids is sparse. In such waters and soils, sensitive organisms such as fish larvae are definitely affected. Certain tree species also show decline if they are simultaneously exposed to both acidic precipitation and toxic gaseous air pollutants such as ozone. In addition, long-term acid rain has been shown to leach away soil buffering capacity.

A second confirmed result of acidic precipitation is the irreversible decay of certain structural materials in buildings, automobiles, statuary, and the like, including many cultural resources. For works of art the damage is of particular concern because the objects are essentially irreplaceable. Vigorous conservation techniques, such as the periodic application of waxes to metal statues, can minimize this problem, but it is difficult to completely prevent degradation from occurring.

Other concerns about acid rain have been proved somewhat less well founded. There is little or no evidence that acid rain is affecting crops or human health, for example, and most lakes seem adequately buffered by their rich organic ecosystems. Nonetheless, the damage that is occurring has been judged of sufficient magnitude to warrant the imposition of emissions controls on acid rain precursors, especially SO_2 and NO_x.

10.11 MAMMALIAN TOXICITY AND THE AUTOMOBILE

Airborne emissions from automobiles are implicated in a number of adverse effects on human health. Photochemical smog (ozone and other products of the smog chemistry) cause a number of respiratory responses including coughing, shortness of breath, and eye, nose, and throat irritation. The effects are particularly severe on children and the elderly. Other respiratory effects are caused by the inhalation of the tiny particles emitted by gasoline and diesel engines. The particles have been linked to increases in hospital visits for emphysema and asthma, and a number of large organic molecules contained in carbonaceous particles are suspected carcinogens.

The incomplete combustion of motor fuels is the source of some 70% of carbon monoxide emissions, and in dense traffic situations the CO levels can be high enough to influence occupational groups such as police officers, parking lot attendants, and toll booth clerks. High CO levels result in adverse impacts on individuals with heart problems, and dizziness, headache, and impaired brain function are potential problems for everyone. Improved exhaust catalysts have gradually decreased CO emissions over the past two decades.

Where leaded gasoline is used, the human health effects are readily measured. The lead leaves the exhaust as a variety of particulate compounds that tend to accumulate by the sides of heavily traveled roadways. This lead can be breathed or eaten, especially by those whose homes or occupations expose them to automotive exhausts on a regular basis. In addition, children in urban areas sometimes consume the soil and the accumulated lead it contains. Most ingested lead is passed from the body, but eventually accumulates if exposure is high enough, causing deterioration or collapse of the central nervous system. Lead is also widely used in wheel balancing weights, but there is no evidence that such use has any adverse effects, since the lead cannot easily be ingested.

Perhaps the most dramatic evidence for improvement in an air quality component as a result of legislation is that for airborne lead in U.S. urban areas. Figure 10.7 shows the 12–yr trend in ambient lead concentrations. It is easy to see that the substantial reductions in leaded gasoline use in the United States that have occurred during this period have been reflected in sharply decreasing atmospheric lead concentrations.

10.12 VISIBILITY AND THE AUTOMOBILE

Visibility is commonly considered to be the greatest distance over which one can see and identify familiar objects with the unaided eye. The concept involves two quite different factors: (1) the degree to which light coming from the object is absorbed or scattered and (2) the visual threshold of perception. In the atmosphere, solar radiation is absorbed and scattered by both gas molecules and particles. The scattering by gases in the clean atmosphere provides an upper limit of about 300 km to the visual range. At moderate or high aerosol loadings, however, radiation scattering by particles is the primary limitation to visibility. On the basis of field measurements, Robert Charlson of the University of Washington has

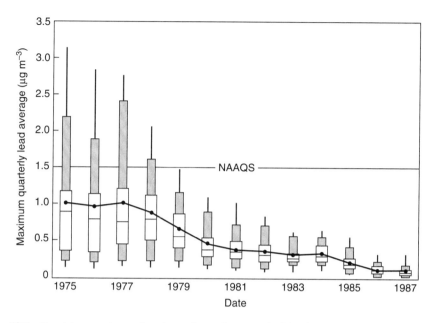

Figure 10.7   Boxplot comparisons of maximum 3–month average lead levels at U.S. sites, 1975–1987. (Environmental Protection Agency, *National Air Quality and Emission Trends Report*, EPA-450/4–84–029, Research Triangle Park, NC, 1985.) The "boxplot" technique that is used to summarize the data plots the average as a dot and the median as a line. The 5th, 10th, and 25th percentiles of the data depict the "cleaner" monitoring sites, the 75th, 90th, and 95th percentiles depict the "dirtier" sites.

shown that the total extinction coefficient ε_{ext} can be related to the total aerosol mass concentration (TPM) by

$$\frac{TPM}{\varepsilon_{ext}} = 2.5 \times 10^5 \ , \tag{10.9}$$

for TPM in units of $\mu g\ m^{-3}$ and ε_{ext} in units of m^{-1}.

In most urban areas, the particles limiting visibility are products of human activity, especially the burning of fossil fuels. Automobile exhaust is the source of only about 2% of all airborne particles, but they tend to be sooty residues with sizes in the range 0. 1–1 μm, just right to absorb solar radiation and inhibit visibility. The result, together with absorption by gaseous NO_2 and light scattering by larger windblown dust particles, is the brownish smoggy air now common throughout the world.

10.13   DESIGN FOR ENVIRONMENT CONSIDERATIONS AT THE PRODUCT USE STAGE

In the product use stage, DFE attempts to eliminate or minimize impacts from solid, liquid, or gaseous residues, consumables, and maintenance materials. A few of the more important guidelines are as follows:

- Optimize product repair and maintenance by improving modularity and access.
- Design systems to work together to achieve efficiency improvements under real-world conditions (e.g., utilize an on-board mapping systems, the Global Positioning System, and smart roadway systems to reduce congestion).
- Utilize recycled material where possible in consumables (oil, antifreeze, etc.).
- Minimize consumables use (sealed bearings, platinum-tipped spark plugs, and longer oil change intervals are examples of how this is being accomplished).
- Package consumable maintenance products with recycled materials and arrange for recycling of the packaging after use.
- Design infrastructure for minimum use of energy and materials, and for maximum containment of solids (e.g., stored road salt), liquids (e.g., gasoline), and gases (e.g., volatiles in paint).

SUGGESTED READING

Beaton, S. P., G. A. Bishop, Y. Zhang, L. L. Ashbaugh, D. R. Lawson, and D. H. Stedman, On-road vehicle emissions: Regulations, costs, and benefits, *Science, 268*, 991–993, 1995.

Calvert, J.G., J.B. Heywood, R.F. Sawyer, and J.H. Seinfeld, Achieving acceptable air quality—Some reflections on controlling vehicle emissions, *Science, 261*, 37–45, 1993.

Committee on Ozone Formation and Measurement, *Rethinking the Ozone Problem in Urban and Regional Air Pollution*, National Academy Press, Washington, D.C., 500 pp., 1991.

Graedel, T.E., and P.C. Crutzen, *Atmosphere, Climate, and Change*, Scientific American Library, New York, 196 pp., 1995.

Lave, L. B., C. T. Hendrickson, and F. C. McMichael, Environmental implications of electric cars, *Science, 268*, 993–995, 1995, and *269*, 741–745, 1995.

Walsh, M., Global trends in motor vehicle use and emissions, *Annual Review of Energy, 15*, 217–243, 1990.

EXERCISES

10.1 Prepare a three page report on the paper by Lave et al. and the letters it inspired (in the Suggested Reading). What are the arguments for and against the use of battery-powered automobiles? Which side, in your view, has the more compelling case? Regardless of whether or not there is a compelling case on either side, what lessons can be drawn from these investigations?

10.2 Prepare a five-page report on one aspect of local scale environmental concerns. The report should have the following outline: explanation of the concern, history of the problem (as supported by data), seriousness of the situation at present, projections for the future, and potential for dealing with the problem. The report may be done for your own local geographical area or for another of your choice. The library and your instructor will be good sources for reference material, in addition to the following. **Smog.** Committee on Tropospheric Ozone Formation and Measurement, *Rethinking the Ozone Problem in Urban and Regional Air Pollution*,

National Academy Press, Washington, DC, 1991; J. H. Seinfeld, Urban air pollution: State of the science, *Science, 243,* 745–752, 1989. **Oil Spills.** National Research Council, *Oil in the Sea: Inputs, Fates, and Effects,* National Academy Press, Washington, D.C., 1985; M. Holloway, Soiled shores, *Scientific American, 265,* (4), 103–116, 1991. **Road Salt.** C.R. Goldman and F.S. Lubnow, Seasonal influence of calcium magnesium acetate on microbial processes in 10 northern Californian lakes, *Resources, Conservation, and Recycling, 7,* 51–67, 1992.

10.3 Prepare a five-page report on one aspect of regional scale environmental concerns, using the outline format of Exercise 10.2. The report may be done for your own regional geographical area or for another of your choice. The library and your instructor will be good sources for reference material, in addition to the following. **Precipitation acidity.** O.P. Bricker and K.C. Rice, Acid rain, in *Annual Review of Earth and Planetary Sciences, 21,* 151–174, 1993; P.M. Irving, Ed., *Acidic Deposition: State of Science and Technology,* US Govt. Printing Office, Washington, DC, 1991; S. E. Schwartz, Acid deposition: Unraveling a regional phenomenon, *Science, 243,* 753–763, 1989. **Visibility.** Committee on Haze in National Parks and Wilderness Areas, *Protecting Visibility in National Parks and Wilderness Areas,* National Academy Press, Washington, DC, 446 pp., 1993; R.B. Husar, J.M. Holloway, D.E. Patterson, and W.E. Wilson, Spatial and temporal pattern of eastern U.S. haziness: A summary, *Atmospheric Environment, 15,* 1919–1928, 1981; P.M. Irving, Ed., *Acidic Deposition: State of Science and Technology,* US Govt. Printing Office, Washington, DC, 1991.

10.4 Prepare a five-page report on one aspect of global scale environmental concerns, using the outline format of Exercise 10.2. The library and your instructor will be good sources for reference material, in addition to the following. **Global climate change.** Intergovernmental Panel on Climate Change, *Climate Change 1995: The Science of Climate Change,* Cambridge University Press, Cambridge, UK, 1996; P.D. Jones and T.M.L. Wigley, Global warming trends, *Scientific American, 263,* (2), 84–91, 1990; B.D. Santer et al., A search for human influences on the thermal structure of the atmosphere, *Nature, 382,* 39–46, 1996. **Ozone depletion.** O.B. Toon and R.P. Turco, Polar stratospheric clouds and ozone depletion, *Scientific American, 264,* (6), 68–74, 1991; S. Solomon, Antarctic ozone: Progress toward a quantitative understanding, *Nature, 347,* 347–354, 1990; **Loss of habitat.** R. Repetto, Deforestation in the tropics, *Scientific American, 262,* (4), 36–42, 1990; J.H. Lawton and R.M. May, Eds., *Extinction Rates,* Oxford, U.K.: Oxford University Press, 233 pp., 1995.

10.5 One of the components of gasoline is heptane, a non-methane hydrocarbon with formula C_7H_{16}. Write and balance the chemical equation for the complete combustion of heptane with molecular oxygen to produce carbon dioxide and water.

10.6 The world's 1990 vehicle fleet of 570 million vehicles emitted about 190 Tg CO. If the world fleet grows by 5% (compounded) per year, and total distance traveled per vehicle remains constant, how much will the CO emission rate have to decrease in order that total automotive CO emissions in 2010 are no higher than they were in 1990?

10.7 The annual average pH of precipitation in New Jersey in 1990 was 4.3. What was the average hydrogen ion concentration in the precipitation?

10.8 The morning rush-hour concentrations of NO_x and NMHC on an average day in Sydney, Australia are 40 ppbv and 1.0 ppmv, respectively. Using Figure 10.5, determine what concentration of ozone would be expected. On a day with little air movement, the same species have concentrations of 100 ppbv and 2.0 ppmv. Under those conditions, what ozone concentration is expected? If emissions evolve so that average morning rush-hour concentrations are 100 ppbv NO_x and 0.6 ppmv NMHC, will ozone concentrations be higher or lower than in the first case above? Why? What if average concentrations become 200 ppbv NO_x and 0.6 ppmv NMHC? Why?

10.9 Within a city square, traffic flow is as follows:

Time	Cars/hr	Trucks/hr
6 AM	1,600	200
7 AM	2,500	350
8 AM	5,600	730
9 AM	8,200	970
10 AM	12,100	1,400
11 AM	15,500	2,300
12 Noon	18,000	3,000

Thirty percent of the vehicles have older emission controls, 70% newer. The square is 0.4 km across, and the average vehicle traverses two sides of the square before leaving it. The average automotive emission factors are CO (old) = 9.4 g km^{-1}; CO(new) = 2.1 g km^{-1}; NO$_x$ (old) = 1.3 g km^{-1}; NO$_x$ (new) = 0.6 g km^{-1}. For trucks, the factors are: CO (old) = 12.5 g km^{-1}; CO (new) = 6.3 g km^{-1}; NO$_x$ (old) = 2.0 g km^{-1}; NO$_x$ (new) = 1.1 g km^{-1}. Compute the emission fluxes of CO and NO$_x$ to the square for each hour between 6 AM and noon.

CHAPTER 11 Design for Recycling

"The concentrations of metal resources in many waste streams that are currently undergoing disposal are higher than for typical virgin resources."

— David Allen, University of Texas

11.1 THE AUTOMOBILE RECYCLING SEQUENCE

The recycling of an automobile occurs in several stages, each stage having its own actors. The sequence is shown in Fig. 11.1. It begins with the transfer to a dismantler of a vehicle deemed no longer suitable for service. The dismantler removes components for which a market exists: usable body panels, the lead-acid battery, wheels and tires, radiator, alternator, and so forth. The remaining vehicle (the hulk) is then sold to the shredder operator.

The shredding operation is accomplished by large machinery that chops the hulk into small pieces 10 cm or so in size and a kilogram or so in weight. These pieces are then sent through a variety of operations that produce three output streams: the "ferrous fraction" (iron, carbon steel, stainless steels), the non-ferrous fraction (aluminum, zinc, copper), and the remainder, termed "automotive shredder residue" (ASR). The ASR is largely metal- and fluid-contaminated polymer materials. Each of the three output streams goes to a further actor in the recycling sequence, the ferrous fraction to a steel mill, the non-ferrous fraction to a non-ferrous separator, where the several metals are separated for resale, and the landfill operator, who receives the valueless ASR. In some countries and in some cases, the ASR undergoes pyrolysis for the recovery of energy.

The automobile recycling system is surprisingly efficient at recovering vehicles at their end of life and reusing at least some of their parts and materials. About 95 percent of

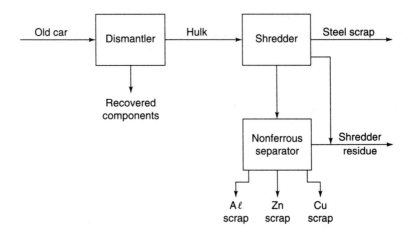

Figure 11.1   The automobile recycling sequence.

all vehicles are eventually involved in this process, compared with an estimated 63 percent of aluminum cans, 30 percent of paper products, 20 percent of glass, and less than 10 percent of plastics. How does the automobile recovery system work, and what are its good and bad points?

11.2 RECYCLING TODAY'S JUNK CARS

The automobile recovery participant with the most intricate organizational operation is the dismantler, whose activities are shown in expanded form in Fig. 11.2. The intricacy is due to the fact that the components the dismantler recovers from an old car are of value to a number of different industries and businesses, and/or that they present difficulties at later vehicle recycling stages. One of the first items recovered is the lead-acid battery. Sometimes the battery itself can be returned to a used-parts market, but generally it is of little value as such and is instead sold to a lead reprocessor, who extracts the lead and resells it to a battery manufacturer. A similar process occurs with the catalytic converter and with electronics components; each go to specialists in the recovery of precious metals or other useful chips or parts, and the recovered materials move on to processors or dealers. Tires that are still road-worthy are reused. If not, they are recycled or incinerated as described in Chapter 10.

Some components of old automobiles can be reused directly, and are immediately sold to spare-parts dealers: wheels, motors for power windows and seats, perhaps the radiator. Others are reusable after reconditioning: alternators, air conditioners, even entire engines. This is particularly important where the engines and components are no longer being manufactured, so recovered and reconditioned parts are the only effective way of keeping older vehicles in service. The wide availability of reconditioned components and the modular design of automobiles are major factors in extending the life of vehicles.

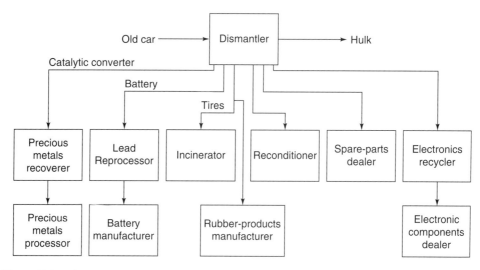

Figure 11.2   Flows of components from automobile dismantlers to component reconditioners, materials reprocessors, and energy recovery operations.

More specialized and more complex parts also see recycling activity. The recent development of automotive electronics has resulted in limited recycling of electronic components, sometimes for the components themselves, sometimes for the precious metals contained in them. The platinum group metals in catalytic converters are quite valuable, and constitute some 30–35% of all the use of these scarce materials. As a result, they have been collected and recycled ever since converters became widespread.

Thus, depending on how far one wants to follow separated metals and resold parts, automobile recycling is a linked activity of between a dozen and two dozen independent participants, each with different roles, different technologies, and different mixes of automotive and non-automotive business. This system has arisen spontaneously as a result of, and is maintained by, economic incentives, not regulatory fiat.

11.3 ECONOMIC ASPECTS OF AUTOMOBILE RECYCLING

A product that is recyclable may or may not be recycled. A distinction here is important: recyclability refers to a product possessing properties such that it is technically possible to recycle it. Recycling is the actual process of recovering materials, components, or other resources (such as energy) from a recyclable product. Recycling is thus strongly related to technology, but, because it does not occur unless profits can be made by the participants, it is also an economic activity. In this way, the automobile differs little from any other complex product, such as a computer or a refrigerator, that is involved in reuse and recycling activities, but that frequently encounters technological or economic barriers to comprehensive recycling.

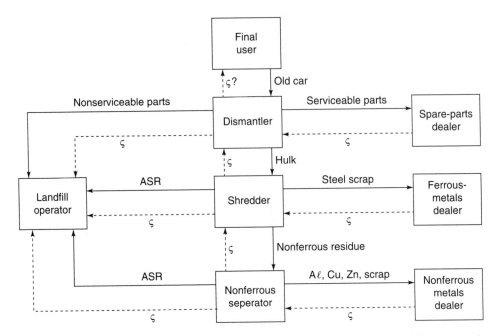

Figure 11.3   Existing flows of materials and payment for them in automobile recycling. ASR is automotive shredder residue and solid lines indicate flows of materials. The non-serviceable parts are those without significant metal content; those that have recoverable metal tend to be left with the hulk. Dashed lines and the generic symbol ζ indicate flows of money. (Adapted from Field, F.R. III, J.R. Ehrenfeld, D. Roos, and J.P. Clark, *Automobile Recycling Policy: Findings and Recommendations*, Cambridge, MA: Center for Technology, Policy, and Industrial Development, Massachusetts Institute of Technology, 1994.)

In Fig. 11.1 and the discussion following, we looked at automobile recycling from an engineering viewpoint. While such information is crucial to informed design and decision making, it is itself incomplete because it does not take into account the economic factors that influence or control many of the interactions. Frank Field of the Massachusetts Institute of Technology has developed a diagram that suggests the interplay between technology and economics in automobile recycling. In his diagram (reproduced in Fig. 11.3), both the materials flows and the economic flows are indicated.

In the initial transfer of materials in Fig. 11.3, that between the last owner of the vehicle and the dismantler, the economic flow may be in either direction, because under different circumstances the dismantler may pay for, or be paid to take, the vehicle. For most of the other transfers, the exchange of resources is made in exchange for payment. The exception is the disposition of material to the landfill, where both the material and payment are transferred from the last owner to the landfill operator.

The automobile recycling process can be shut down at any step in the flow of materials if either the technology or the economics is unsatisfactory. For example, the shredder operator sells scrap steel to steel mills, which recycle it. In the 1960s, most steel mills were of the open hearth variety, which could produce steel with 40–45% scrap in the charge. In the 1970s, many steel makers switched to a new and more efficient technology, the basic oxygen furnace, which could use only 25–20% scrap. The result was a lowering in demand for scrap steel and a rapid accumulation of junked automobiles that were economically unattractive to recycle. The situation changed again in the 1980s, as another new steelmaking technology, the electric arc furnace, came into being. This new process could produce usable steel from charges of 80% scrap steel or higher. Suddenly, the recycling of automobiles became much more economically advantageous.

What are the messages to be gained from consideration of Fig. 11.3? One was mentioned earlier: that the recycling infrastructure is complicated, and has many actors. A further complication, not shown in the figure, is that most of the actors in automotive recycling are involved in the recycling of other products as well: washing machines, plumbing fittings, electronic devices, and so on. If appliance recycling is a recycler's major business and automotive recycling a sideline, as could certainly be the case for non-ferrous scrap metal dealers or plastics reprocessors, his corporate viability might be influenced in only a minor way by mandated actions dealing solely with automotive material or automotive recycling economics. Small increases in the cost of automotive scrap resulting from ill-considered regulatory intervention could backfire; history shows that the effects of reporting requirements on delivery of auto hulks, or subsidized payments for unwanted materials, are to produce large bureaucracies, abandoned vehicles, and black markets. Accordingly, public policies intended to increase automobile recycling are unlikely to be successful if they are directed at influencing market transactions. More successful policies are likely to be those that aim at creating incentives for increased technological innovation by the private firms themselves: increasing the efficiency of a recycling process or improving the quality of the recovered material, for example.

11.4 RECYCLING TODAY'S AUTOMOTIVE INFRASTRUCTURE

In general, the rule for infrastructure is no different than that for any other product: all infrastructure should be designed *ab initio* to be modular and recyclable at end of life. Horizontal recycling is preferable to vertical recycling, but the economics for this may be difficult given the low cost of much of the virgin material used for roads and highways. As with the automobile, the recycling of the automotive infrastructure occurs in several stages, each stage with its own actors. The sequence is shown in Fig. 11.4.

The infrastructure material recycled in the largest amounts is asphalt pavement. This process involves pavement fragmentation, followed by rotor-milling to reduce the pavement chunks to very nearly the size of the original aggregate. Depending on the availability of equipment and the ease of transportation, the milling may be done at the construction site or the paving fragments may be taken to a rotor-mill at a fixed site. The resulting milled material is then reheated, reformulated with the addition as necessary of new bitumen, and reapplied to the base surface material. It is often quicker to recycle pavement on-site than to import and use virgin materials, a characteristic that permits minimum disruption of traffic flow during repaving activities.

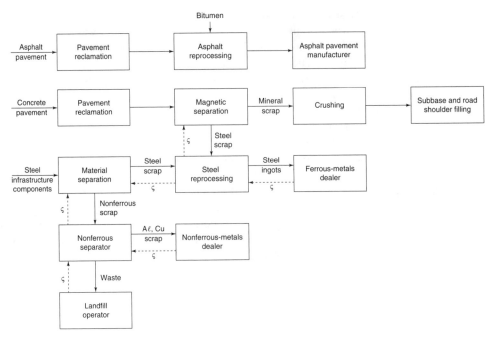

Figure 11.4   The automotive infrastructure recycling sequence. Solid lines indicate flows of materials. Dashed lines and the generic symbol ζ indicate flows of money.

The potential for recycling of asphalt pavement can be high if the proper materials were used in the pavement in the first place. The key ingredient is the aggregate, which must be non-porous and have good integrity so that it will retain a rough surface and contribute to a road surface with good traction. Such aggregates are said to resist "polish." Where attention has been given to incorporating high-quality aggregate, and where the rather expensive rotor-milling equipment is available, recycling rates of 90% or higher are not unusual. This is particularly true in urban areas. Overall, the asphalt recycling rate is probably in the neighborhood of 50% and is growing.

Concrete undergoes end-of-life processing by crushing, followed by magnetic separation to separate the steel reinforcing bar and rod fragments from the mineral matrix. These metal fragments are then recycled as with any relatively impure industrial steel. The mineral-cement matrix has several potential uses. The best, if energy costs do not make it prohibitive, is sending lightly-crushed concrete to an aggregate supplier for final crushing. The resulting material is then used just as is granular material from natural sources. Alternatively, the crushed matrix material can be used as filler for highway construction or modification.

Infrastructure components can be recycled as can any large industrial product. Among current practices in a number of locations are the reuse of aluminum signs following stripping and cleaning (at one-third to one-fifth the cost of new signs), the reconstruction of metal-beam guardrail, and the reuse at the same or different locations of hardware such as manhole covers and frames, lighting standards, and chain-link fencing. For hardware components, especially, design for environment approaches can be helpful in making recycling easier and more profitable.

11.5 DESIGN FOR RECYCLING

Today's recyclers deal with vehicles and infrastructure components that were designed without the slightest thought as to their recyclability. As a consequence, many of the materials are difficult to recover in pure form, many components are difficult to reuse, many parts of the vehicle are hard to separate. Major improvements are possible if design engineers think from the beginning about designing for recycling.

When planning for product end-of-life, two complementary types of recycling should be considered, as discussed in Chapter 7. One, previously termed *horizontal recycling*, can also be designated *closed-loop recycling*. As seen in Fig. 11.5, closed-loop recycling involves reuse of the materials to make the same product over again. (A typical example is reprocessing used aluminum cans to make new aluminum cans.) The alternative is *open-loop recycling*, also termed *vertical recycling* or *cascade recycling*), which reuses materials to produce different products. (A typical example is using discarded office paper to make brown paper bags.) The mode of recycling will depend on the materials and products involved, but closed-loop cycling should generally be preferred if possible.

Design for recycling (DFR) should focus on a small number of rules:

- Minimize the use of materials
- Minimize materials diversity, i.e., the number of different materials used
- Choose desirable materials, considering not just manufacturing and use characteristics, but recycling potential as well
- Make it modular
- Eliminate unnecessary product complexity
- Make it efficient to disassemble
- Make the materials easy to recover

As with most aspects of engineering, lessons learned through experience and experiments have provided much guidance in these areas. Let us examine them one by one.

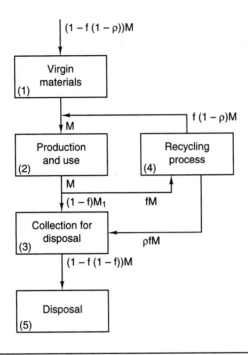

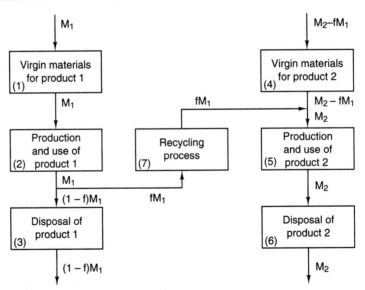

Figure 11.5   Closed-loop (top) and open-loop (bottom) recycling of materials. In the diagrams, M refers to mass flows, f to the fraction of the flow delivered to the recycling process, and ρ to the fraction of that flow rejected as unsuitable for recycling. (B.W. Vigon, D.A. Tolle, B.W. Cornaby, H.C. Latham, C.L. Harrison, T.L. Boguski, R.G. Hunt, and J.D. Sellers, *Life-Cycle Assessment: Inventory Guidelines and Principles*, EPA/600/R-92/036, US Environmental Protection Agency, Cincinnati, OH, 1992.)

11.5.1   Minimize the Use of Materials

In a "less is better" philosophy, design goals should be accomplished by the clever use of minimal amounts of materials. The strength desired in a component or panel can be achieved with ribs, bosses, and the like rather than heavy-gauge materials. Advanced materials with improved structural characteristics can provide the desired degree of stiffening with less material. Detailed stress computations may demonstrate that less material is needed than has customarily been used.

11.5.2   Minimize Materials Diversity

Any article as complex as an automobile inevitably requires the use of many different materials. Frequently, however, material selection is not optimized across the automobile design as a whole, with the result that a greater variety of materials than cost and performance criteria actually require are used. The result is a product that may be unnecessarily expensive to recycle or, if its residues are chemically incompatible, actually may become more difficult to recycle from a technical standpoint.

11.5.3   Choose Desirable Materials

Automotive materials have always been chosen with their performance characteristics in mind, but the environmental aspects of materials can enter into the choices as well. A key recommendation is to use recycled materials whenever possible. Recycled aluminum and steel are obvious choices, provided their materials specifications are satisfactory, but recycled plastics and filler materials of various kinds are also possible. And, regardless of whether the materials used are virgin or recycled, their use should be such that their recycling at end of life is optimized.

Where they can be used for automotive purposes, biomaterials (for example, wood or flax) have much to recommend them. Many, for example, have excellent mechanical properties at modest weight. Biomaterials use does not deplete resources and avoids toxicity problems. Further, when they eventually decompose or undergo incineration, they return to the atmosphere no more carbon dioxide than they absorbed from the atmosphere while growing. The agricultural operations that produce them have their own environmental impacts, so tradeoffs are involved. Among the uses to which biomaterials are being put in automobiles are shelves, floor mats, panels, and interior padding.

11.5.4   Make It Modular

Modularity has always been a feature of automobiles. Shock absorbers, radiators, exhaust systems, even engines are commonly replaced. Designers can aid in this process by designing for efficient replacement of modules—the use of standard sizes and types of fasteners, for example—and by designing modules so that they may be efficiently recycled, such as by making fluid-containing parts easy to drain and clean.

Manufacturers can readily aid this process by taking a few logical steps now generally inhibited by tradition. Mercedes-Benz in Germany, for example, works with repair and collision shops to recover modules for reconditioning or remanufacturing. If neither is possible, the materials in the modules are recycled. In the longer term, it may be possible for reconditioned or remanufactured components to be given standard "new parts" guarantees and used in new automobiles. A next step in modular design is the creation of systems that permit the graceful upgrade of individual modules while the rest of the automobile remains the same. One might imagine, for example, exchanging control panel and engine modules while retaining the passenger compartment and its heating and cooling system. A few preliminary modular designs suggest that such vehicles may be available in the early 21st century.

11.5.5   Eliminate Unnecessary Complexity

It is well known within industry that far more individual parts than necessary are commonly incorporated into product designs. This not only makes manufacture more complex and expensive, but environmentally difficult as well. More parts require more cleaning, which in turn generates more waste streams. Parts complexity can have a similar effect on the recycling end: a profusion of parts makes disassembly more complex, and tends to encourage the use of generic end-of-life technologies such as hammer mills and shredders, which produce larger volumes of ASR than necessary.

11.5.6   Make It Efficient to Disassemble

A major factor in recyclability is how easy or difficult it is for a vehicle to be disassembled. Where once parts of vehicles were welded or otherwise joined in ways difficult to reverse, modern fastening technology provides many joining alternatives, should they be exercised. For example, joining parts with snaps, clamps, or screws is preferable to using welds or glues. Bolts and screws should be positioned so that access to them is relatively easy, and the fasteners should be those in common use, recognizing that dismantlers are likely to have on hand only the more common tools.

Fastening techniques thought to be relatively unconventional are increasingly becoming more common. For example, polymeric hook and loop fasteners are used by some manufacturers to affix head linings and interior trim into place. Hook and loop fasteners become only more secure as vehicles flex during use, but the components they join can be readily separated at recycling time.

Once materials are disjoined, it is crucial to be able to identify them promptly and reliably. In this connection, standard identification markings, such as those for plastics of the International Organization for Standardization (ISO), should always be used (Fig. 11.6). Marking is also useful for metals, should there be any uncertainty about the metal or alloy from which a component is made.

Since trace impurities can affect the value of scrap materials, and hence their recyclability, the thoughtful designer will try to make trace materials easy to separate. Copper wiring harnesses, for example, should be easy to strip from auto hulks, thus avoiding the

>PC<	Poly (carbonate)
>PBT<	Poly (butylene terephthalate)
>(PBT + PC)<	Poly (butylene terephthalate)/poly(carbonate) blend
>(PBT+PC) – GF20<	Poly (butylene terephthalate)/poly(carbonate) blend; 20% glass-filled

Figure 11.6   Examples of standardized markings for plastic parts. The complete specifications are in documents ISO/DIS 11469, *Generic Identification and Marking of Plastic Parts*, and ISO 4043, *Plastic Symbols, Parts 1–3*, International Organization for Standardization, Geneva.

contamination of the steel. Natural materials such as wood or flax should be easy to separate from plastics or metals.

11.5.7   Make the Materials Easy to Recover

A major impediment to the recycling of automotive materials is their presence in composites or welded or glued units that make the individual materials difficult to recover in pure form. There are two factors at work here. One is an initial assemblage of materials that constrains recovery, such as carbon fibers in a polymer matrix, or a wood, metal, and polymer mixture in a dashboard. Mixed materials are also a problem with plastics, some combinations of which are compatible during recycling, some not. A second problem is a difficulty in separating otherwise relatively pure materials, such as the copper in a wiring harness buried within door panels. In either case, one or all of the mixed materials are unlikely to be economically recoverable, given modern automotive designs. Copper is a particular problem if not retrieved, since copper impurities inhibit the mechanical properties and hence the reuse of recycled steel. Although aluminum is a less efficient electrical conductor, it may be a suitable replacement for copper in some current-carrying applications, and fibre optics may be suitable where information rather than current is being transmitted.

A related problem is that of inserts, that is, components that are joined mechanically, such as in metal studs inserted into plastic components. Designs in which this situation occurs are generally unsound from a DFE standpoint; if they must be used, the insert should generally be of steel, which can be separated magnetically with some effort.

Coatings and platings are an obvious example of the mixing of materials. It is often the case that during recycling such surface treatments are lost to society, as there is no reasonable means of recovery. For automobiles, this is particularly true of the zinc used as an anticorrosion plating. Careful materials design can sometimes enable compatible coatings to be used for plastics, and metal platings can sometimes be recovered, but, in general, painting or plating, especially with toxic substances such as chromium, is to be avoided.

Fluid recovery is also related to the design process. Although recovery of oil, antifreeze, transmission fluid, and the like is generally practiced, good design can improve the completeness of the recovery process. Drainage points should be located and fluid reservoirs designed so as to be completely drainable, for example, and drainage plugs and access ports standardized as much as possible.

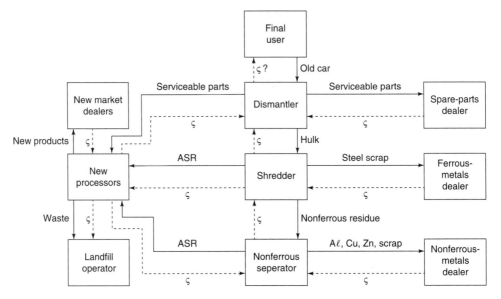

Figure 11.7   Potential flows of materials and payment for them in an expanded automobile resource recovery scenario. ASR is automotive shredder residue and solid lines indicate flows of materials. Dashed lines and the generic symbol ζ indicate flows of money. (After Field, F.R. III, J.R. Ehrenfeld, D. Roos, and J.P. Clark, *Automobile Recycling Policy: Findings and Recommendations*, Cambridge, MA: Center for Technology, Policy, and Industrial Development, Massachusetts Institute of Technology, 1994.)

A final point relates to designs that are presumably without mixed materials but which achieve mixing during manufacture. The classic case is that of labels affixed to plastic parts to provide bar codes, mandated consumer information, safety instructions, and the like. Very often, these labels are difficult and time-consuming to remove, and capable of irretrievably contaminating the basic material if left in place. The solution is to make the labels easily strippable, to make the labels from the same plastic as the part itself, or to fasten them onto a small portion of the part that is designed to be broken off and discarded at the recycling stage.

11.6 FUTURE RECYCLING POSSIBILITIES

Earlier in this chapter, we presented information on the flows of materials and economics in recycling. How might we wish that diagram to change in the future? One perspective is shown in Fig. 11.7 for the case of the automobile. Superficially, the difference between Figures 11.3 and 11.7 is that residues formally regarded as wastes are now sent to a new group of processors who are equipped to turn the bulk of those residues into materials or products with satisfactory market value. A second, implied, difference between the figures

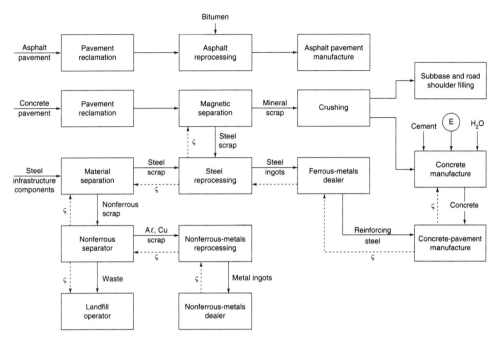

Figure 11.8   Potential flows of materials and payment for them in an expanded automotive infrastructure resource recovery scenario. Solid lines indicate flows of materials. Dashed lines and the generic symbol ζ indicate flows of money.

is that the original processes themselves have become more efficient and thus more profitable, both for the operators and for society as a whole.

As with the automobile, we can construct a diagram for how one might wish the infrastructure recycling to look in the future; it appears in Fig. 11.8. The difference between that figure and Fig. 11.4 is not great, partly because the asphalt recycling system at present is reasonably efficient and partly because roadway contractors perform many of the process steps themselves on site rather than interacting with suppliers to the degree that automobile manufacturers do. The principal addition that appears on Fig. 11.8 is the horizontal recycling of concrete, a process that does not now exist. Should it be developed, one can imagine that the volume of construction debris being discarded, often now 20–25% of total input to landfills, could be markedly decreased.

What are the prospects for these changes? Without question, they are being enhanced by the formation of recyling consortia, in which automobile and infrastructure manufacturers work together to improve dismantling and materials sorting techniques, to identify areas in which research is needed to surmount current recycling constraints, evaluate environmental impacts, and promote markets for recycled materials. Consortia now exist in Germany, Scandinavia, and North America, with globalization of the recycling effort a potential future activity. Among the areas where effort can already be seen to be needed are better parts recovery and reuse, enhanced selection and eventual reuse of plastics, and reclamation of

ASR. Another need is a change in "mind set" by the materials suppliers such that the materials processing capacity currently consumed by the production of virgin materials also be used for reprocessing scrap material. In many cases, the processing of postconsumer materials is cheaper, faster, and cleaner than the original process.

The ultimate concept in automotive design for environment has been conceived by Mercedes-Benz. Its centerpiece is a "universal chassis" on which can be mounted any of a family of vehicle bodies and interiors. Originally, a young adult might purchase the chassis and a sporty car body. A few years later, when the person married, the sporty car body would be exchanged for a small sedan. Still later, with the arrival of children, a station wagon or minivan body is substituted. The final phase might be trading the wagon or minivan body for a midsize sedan. If the chassis were made sufficiently rugged and long-lived, this approach would conserve materials while providing for the changing needs of vehicle owners. The concept obviously involves substantial engineering challenges in reliability, maintenance, modular design, and upgrading possibilities, but seems a useful goal to have in the back of one's mind as tomorrow's transportation systems are developed.

SUGGESTED READING

Ahmed, I., *Use of Waste Materials in Highway Construction*, Park Ridge, N.J.: Noyes Data Corp., 114 pp., 1993.

Field, F.R. III, J.R. Ehrenfeld, D. Roos, and J.P. Clark, *Automobile Recycling Policy: Findings and Recommendations*, Cambridge, MA: Center for Technology, Policy, and Industrial Development, Massachusetts Institute of Technology, 1994.

Henstock, M.E., *Design for Recyclability*, London: Institute of Metals, 135 pp., 1988.

Heyl, F.D., and G.P. Jayaprakash, Eds., From refuse to reuse: Recyling and reuse of waste materials and by-products in transportation infrastructure, *TR News*, No. 184, Washington, D.C.: Transportation Research Board, 1996.

McPhee, J., Duty of care, *The New Yorker*, pp. 72–80, June 28, 1993.

EXERCISES

11.1 One of the structural problems with the automobile recycling system is that enhanced recyclability designed into a product by the manufacturer benefits the recycler, not the manufacturer. Incentives for DFR are therefore weakened. (a) Suggest three policies to overcome this disincentive. (b) Do any of your policies raise potential legal concerns, such as anti-trust problems?

11.2 Social and legal pressure encourages automobile manufacturers to "lightweight" their vehicles so as to use less fuel. A principal means of accomplishing this is to substitute lighter materials, such as plastic or aluminum, for steel. At some point, however, the lowered steel percentage in the car body may make scrap steel recovery uneconomic. Given the industry structure shown in Fig. 11.3, what would the implications of this situation be? What policies might you suggest to balance the interests in fuel efficiency and maximizing recycling? What data might you want before deciding on a policy course?

11.3 In a closed-loop recycling system, the mass flow M is 5000 kg/hr, f is 0.7, and ρ is 0.1. Diagram the system and indicate all flow rates on the diagram.

11.4 In an open-loop recycling system, the mass flows M_1 and M_2 are 8000 kg/hr and 3000 kg/hr and f is 0.6. Diagram the system and indicate all flow rates on the diagram.

11.5 In the open-loop system of problem 11.4, assume that the recyling process rejects 15% of the material provided to it. Diagram this altered system and indicate on the diagram all the flow rates.

11.6 The average 1990s automobile contains 65 kg of aluminum. This material can be supplied by mining and electrolytically processing bauxite [$Al(OH)_3$] at an energy cost of 270 GJ/Mg or by recycling aluminum from cans and other sources at an energy cost of 17 GJ/Mg. Compute the energy consumed in supplying the aluminum (a) from virgin materials, and (b) from recycled materials, for the 49 million new vehicles expected to be built this year.

11.7 Most of the world's automobiles are built in Europe, Asia, and North America, the approximate percentages being 37, 28, and 31. The total annual energy use in these regions for all purposes (1990 figures) are 90, 45, and 97 Quads. Using the results of Exercise 11.6, what percentage of total energy use would be consumed in these regions if the aluminum in the automobiles were provided from (a) virgin materials; (b) recycled materials? What is not taken into account in this exercise?

CHAPTER 12

How Green Is the Automobile and Its Infrastructure?

"What gets measured gets managed." — Business School Aphorism

12.1 LIFE CYCLE ASSESSMENT

A primary thrust of industrial ecology is that manufacturers practice product stewardship—designing, building, maintaining, and recycling products in such a way that they pose minimal impact to the wider world. Product stewardship should be broadly interpreted to include services, which should also be performed so as to have minimal impact. The way in which these tasks are addressed in a formal manner is by the process of life cycle assessment (LCA), a family of methods for looking at materials, services, products, processes, and technologies over their entire life.

The essence of life cycle assessment is the evaluation of the relevant environmental, economic and technological implications of a material, process, or product across its life span from creation to waste or, preferably, to re-creation in the same or another useful form. The Society of Environmental Toxicology and Chemistry defines the LCA process as follows:

> "The life-cycle assessment is an objective process to evaluate the environmental burdens associated with a product, process, or activity by identifying and quantifying energy and material usage and environmental releases, to assess the impact of those energy and material uses and releases on the environment, and to evaluate and implement opportunities to effect environmental improvements. The assessment includes the entire life cycle of the product, process or activity, encompassing extracting and processing raw materials; manufacturing, transportation, and distribution; use/re-use/maintenance; recycling; and final disposal."

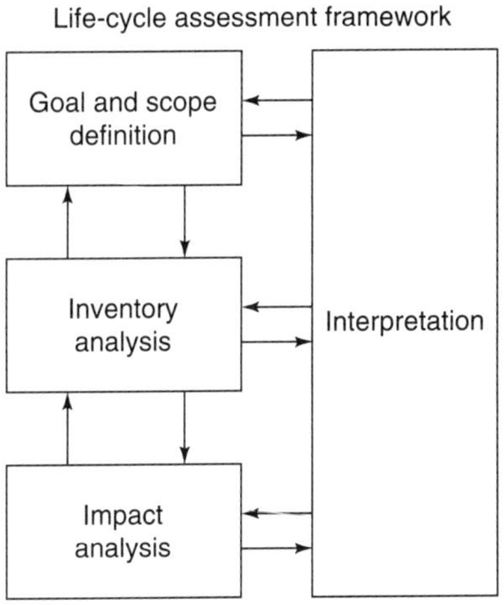

Life-cycle assessment framework

Figure 12.1   Components in the life cycle assessment of a product. (Adapted from Technical Committee 207, International Organization for Standardization, Geneva, 1996.)

Such an analysis is a large and complex effort, and there are many variations. Nonetheless, there is preliminary agreement on the formal structure of LCA, which contains four components: *Goal and scope definition*, *inventory analysis*, *impact analysis*, and *interpretation*. The concept of the life cycle methodology is pictured in Fig. 12.1. First, the goal and scope of the LCA is defined. An inventory analysis is then performed. The results of this analysis can sometimes be interpreted to provide guidance on environmentally preferable actions, but it is more likely that guidance will be provided after the impact analysis is conducted. The potential improvements can then be evaluated for feasibility, cost, and other non-environmental factors. Those that are adopted probably influence the inventory analysis, so iteration of the LCA steps is sometimes involved. Following the adoption of the changes encouraged by the LCA, the improved product design is released for manufacture.

The second component of LCA, inventory analysis, is by far the best developed. It uses quantitative data to establish the levels and types of energy and materials inputs to an industrial system and the environmental releases that result, as shown schematically in Fig. 12.2. Note that the approach is based on the idea of a family of materials budgets, measuring the inputs of energy and resources that are supplied and the resulting products, including those with value and those that are potential liabilities. The assessment is done over the entire life cycle—materials extraction, manufacture, distribution, use, and disposal.

The third stage in LCA, the impact analysis, involves relating the outputs of the system to the impacts on the external world into which those outputs flow. We present aspects of this difficult and potentially contentious stage later in this book. The final stage, interpretation, is the explication of needs and opportunites for reducing environmental impacts as a result of industrial activity being performed or contemplated. It follows directly from the completion of stages one to three, and in implementation is termed "Design for Environment"; many of its tenets have been discussed earlier in this book.

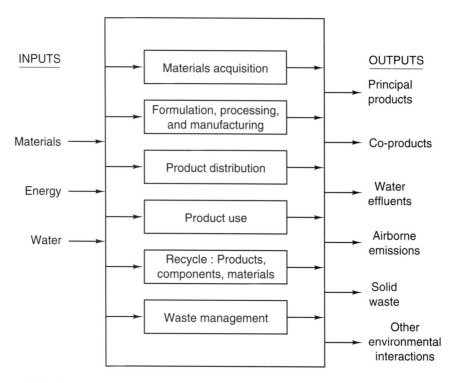

Figure 12.2   The elements of a life cycle inventory analysis. (Adapted from Society of Environmental Toxicology and Chemistry (SETAC), *A Technical Framework for Life-Cycle Assessment*, Washington, D. C., 1991.)

All LCA methodologies are in their infancy. As more experience is gained and more products, processes, and materials subject to assessment, the LCA approaches will become more useful and more efficient. It is unlikely that one single methodology will be optimal for all DFE analyses. Packaging, bulk chemical products, consumer care products, food products, complex manufactured items with relatively long lives (airplanes) as opposed to short lives (radios, telephones)—all have quite different characteristics, maintenance needs, life cycles, and environmental impacts. It is reasonable to assume that the tools of LCA evaluation will become more sophisticated with time.

12.2   THE IVL/VOLVO EPS SYSTEM FOR LIFE-CYCLE ASSESSMENT

To address life-cycle assessment for the automobile in a formal way, the Swedish Environmental Institute (IVL) and the Volvo Car Corporation have developed an analytic tool designated the Environment Priority Strategies for Product Design (EPS) system. The goal of the EPS system is to allow product designers to select components and subassemblies that minimize environmental impact. Analytically, the EPS system is quite straightforward, though detailed. An environmental index is assigned to each type of material used in automobile manufacture. Different components of the index account for the environmental impact of this material during product manufacture, use, and disposal. The three life cycle stage components are summed to obtain the overall index for a material in "environmental load units (ELUs)" per kilogram (ELU/kg) of material used. The units may vary. For example, the index for a paint used on the car's exterior would be expressed in ELU/m^2.

When calculating the components of the environmental index, the following factors are included:

- Scope: A measure of the general environmental impact.
- Distribution: The size or composition of the affected area.
- Frequency or Intensity: Extent of the impact in the affected area.
- Durability: Persistence of the impact.
- Contribution: Significance of impact from 1 kg of material in relation to total effect.
- Remediability: Cost to remediate impact from 1 kg of material.

These factors are calculated by a team of environmental scientists, ecologists, and materials specialists to obtain environmental indices for every applicable raw material and energy source (with their associated pollutant emissions). A selection of the results is given in Table 12.1. A few features of this table are of particular interest. One is the very high values for platinum and rhenium in the raw materials listings. These result from the extreme scarcity of these two metals. The use of CFC-11 is given a high environmental index because of its effects on stratospheric ozone and global warming. Finally, the assumption is made that the metals are emitted in a mobilizable form. To the extent that an inert form is emitted, the environmental index may need to be revised.

Table 12.1   A Selection of Environmental Indices (units: ELU/kg)*

Raw Materials:		Emissions-Air:	
Co	76	CO_2	0.09
Cr	8.8	CO	0.27
Fe	0.09	NO_x	0.22
Mn	0.97	N_2O	7.0
Mo	1.5E3	SO_x	0.10
Ni	24.3	CFC-11	300
Pb	180	CH_4	1.0
Pt	3.5E5		
Rh	1.8E6	Emissions-Water:	
Sn	1.2E3	Nitrogen	0.1
V	12	Phosphorus	0.3

*Steen, B., and S. Ryding, *The EPA Enviro-Accounting Method: An Application of Environmental Accounting Principles for Evaluation and Valuation of Environmental Impact in Product Design*, Swedish Environmental Research Institute (IVL), Stockholm, Sweden, 1992. $1.5E3 = 1.5 \times 10^3$.

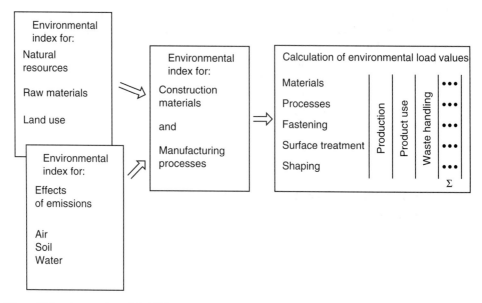

Figure 12.3   An overview of the EPS system, showing how the calculation of summed environmental load values proceeds. (Reproduced with permission from B. Steen, and S. Ryding, *The EPA Enviro-Accounting Method: An Application of Environmental Accounting Principles for Evaluation and Valuation of Environmental Impact in Product Design.* Copyright 1992 by Swedish Environmental Research Institute (IVL).)

Given an agreed set of environmental indices, they are multiplied by the materials uses and process parameters (in the appropriate proportions) to obtain environmental load units for processes and finished products. The entire procedure is schematized in Fig. 12.3. Note that it includes the effects of materials extraction, emissions, and the impacts of manufacturing, and follows the product in its various aspects through the entire life cycle.

As an example of the use of the EPS system, consider the problem of choosing the more environmentally responsible material to use in fabricating the front end of an automobile. As shown in Fig. 12.4, two options are available: galvanized steel and polymer composite (GMT). They are assumed to be of comparable durability, though differing durabilities could potentially be incorporated into EPS.

Based on the amount of each material required, the environmental indices are used to calculate the environmental load values at each stage of the product life cycle. Table 12.2 illustrates the total life cycle ELU's for the two front ends. All environmental impacts, from the energy required to produce a material to the energy recovered from incineration or reuse at the end of product life, are incorporated into the ELU calculation. To put the table in the LCA perspective, the kg columns are LCA stage one, the ELU/kg columns are the environmental indices, and the ELU columns are LCA stage two. There are several features of interest in the results. One is that the steel unit has a larger materials impact during manufacturing, but is so conveniently reusable that its overall ELU on a materials basis is lower than that of the composite. However, it is much heavier than the composite unit, and that factor results in much higher environmental loads during product use. The overall result is one that was not intuitively obvious: that the polymer composite front end is the

**Which front end is more
environmentally sound?**

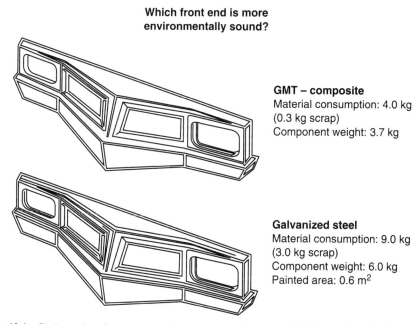

GMT – composite
Material consumption: 4.0 kg
(0.3 kg scrap)
Component weight: 3.7 kg

Galvanized steel
Material consumption: 9.0 kg
(3.0 kg scrap)
Component weight: 6.0 kg
Painted area: 0.6 m^2

Figure 12.4   Design options for automotive front end pieces. (Courtesy of I. Horkeby, Volvo Car Corporation.)

better choice in terms of environmental impacts during manufacture, the steel unit the better choice in terms of recyclability, and the polymer composite unit the better overall choice because of lower impacts during product use. Attempting to make the decision on the basis of an analysis of only part of the product life cycle would result in an incompletely guided and potentially incorrect decision.

Another interesting aspect of automotive components revealed by EPS analysis is that different parts of the automobile have their highest environmental impacts at different life stages. As shown in Fig. 12.5, automotive electrical systems have their highest impacts at the production life stage; the in-use impacts are quite small, while those at end of life are significant but much lower than during manufacture. In the case of seat covers, the production and in-use stage impacts are nearly equal, while the end of life stage impact is small. Both of these differ from the front end, where the in-use impacts are far greater than those at any other life stage.

All quantitative LCA systems, including EPS, must make assumptions about how long a product will be in service and how it will be used. Consider, for example, the automotive component termed a *bonnet* in Europe and a *hood* in North America. The EPS assessment of environmental loads during manufacture favors the steel bonnet (Fig. 12.6), largely because of the substantial amount of energy needed to process aluminum from its ore or recycle streams. The lighter aluminum bonnet requires less fuel consumption and is thus responsible for lower exhaust emissions, however. Because of these characteristics, the calculated environmental loads of the two materials are equal at about 15,000 km of vehicle distance traveled. At higher use levels, the aluminum bonnet becomes increasingly desirable.

Table 12.2   Calculation of Environmental Load Values for Automobile Front Ends*

Materials & Processes	Production			Product Use‡			Waste						Total
							Incineration			Reuse			
	ELU/kg	kg	ELU	ELU/kg	kg	ELU	ELU/kg	kg	ELU	ELU/kg	kg	ELU	ELU
GMT-composite													
Production:													
GMT material	0.58	4.0	2.32										2.32
Reused pro-duction scrap	−0.58	0.3	−0.17										−0.17
Compression molding	0.03	4.0	0.12										0.12
Product Use:													
Petrol				0.82	29.6	24.27							24.27
Recycling:													
GMT material							−0.21	3.7	−0.78				−0.78
Total sum			2.27			24.27			−0.78				25.76
Galvanized steel													
Production:													
Steel material	0.98	9.0	8.82										8.82
Steel stamping	0.06	9.0	0.54										0.54
Reused pro-duction scrap	−0.92	3.0	−2.76										−2.76
Spot welding (spots)	0.004	48	0.19										0.19
Painting (m²)	0.01	0.6	0.02										0.02
Product Use:													
Petrol				0.82	48.0	39.36							30.72
Recycling:													
Steel material										−0.92	6.0	−5.52	−5.52
Total sum			6.81			39.36						−5.52	40.65

*Ryding, S., B. Steen, A. Wenblad, and R. Karlsson, *The EPS System—A Life Cycle Assessment Concept for Cleaner Technology and Product Development Strategies, and Design for the Environment*, paper presented at EPA Workshop on Identifying a Framework for Human Health and Environmental Risk Ranking, Washington, DC, June 30—July 1, 1993.

‡ The ELU/kg figure is based on one year of use. For the automobile, an eight-year life is assumed, hence the second product use entry is eight times the actual weight.

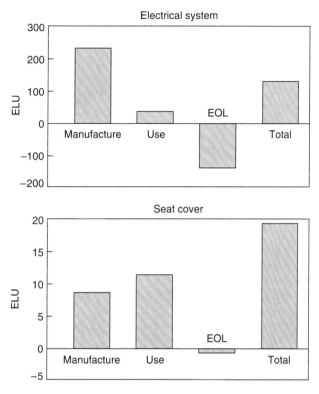

Figure 12.5   Environmental impacts (in ELU) of automotive electrical systems and seat covers as a function of life-cycle stage (EOL. = end of life). (Courtesy of A. Wendel, Volvo Car Corporation.)

In principle, an entire automobile can be assessed by summing the assessments of the individual components: front end, electrical system, seats, bonnet, frame, and so forth. It is not clear at present that such an operation serves a truly useful environmental purpose. Does a vehicle with a summed ELU of, say, 40,500 really represent a greener product than one with a summed ELU of 42,200? Whether it does or not, does the overall assessment provide useful guidance to the design team? Most importantly, summed ELUs assuredly do not capture the extraordinarily complex systems impacts of automobiles that have been discussed throughout this text. It seems probable that the real value of the EPS assessment is in forcing designers of individual components to look hard at the environmental impacts of their components at different stages of the life cycle, not in providing justification for the marketing of environmentally preferable automobiles.

The EPS system is currently being refined and implemented by several organizations affiliated with IVL. Corporations in Sweden and elsewhere have expressed great interest in developing EPS systems that are specific to their products and manufacturing procedures. The EPS's greatest strength is its flexibility; raw materials, processes, and energy uses can be added easily. If a manufacturing process becomes more efficient, all products that utilize

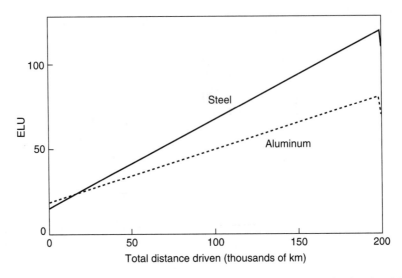

Figure 12.6   The accumulated environmental loads (in ELU) for automobile bonnets of 15 kg of steel (solid line) and 9 kg of aluminum (dashed line) as a function of total distance driven. (Courtesy of A. Wendel, Volvo Car Corporation.)

the process will automatically possess upgraded ELUs. Perhaps the EPS System's greatest weakness is its need to quantify uncertain data and compare unlike risks, in the process making assumptions that gloss over serious value and equity issues. The problems are endemic, however, to any system that aggregates LCA data as the EPS system does.

12.3 STREAMLINED PRODUCT ASSESSMENT

Often the information needed by a designer in order to do a full LCA of the type described above may not be available, or time or other resources may be too short. In such a situation, must LCA be abandoned? Faced with limitations but interested in improving environmental performance, industries in the past several years have developed a number of techniques for producing streamlined versions of LCAs. A suitable streamlined assessment system should have the following characteristics: it should lend itself to direct comparisons among rated products, be usable and consistent across different assessment teams, encompass all stages of product life cycles and all relevant environmental concerns, and be simple enough to permit relatively quick and inexpensive assessments to be made. Clearly, it must explicitly treat the five life-cycle stages in a typical complex manufactured product. Stage 1, premanufacturing, is performed by suppliers, drawing on (generally) virgin resources and producing materials and components. Stage 2, the manufacturing operation, and Stage 3, packaging and shipping, are directly under corporate control. Stage 4, the customer use stage, is not directly controlled by the manufacturer, but is influenced by how products are designed and by the degree of continuing manufacturer interaction. In Stage 5, a product no

Table 12.3   The Environmentally-Responsible Product Assessment Matrix

Life Stage	Environmental Concern				
	Materials Choice	Energy Use	Solid Residues	Liquid Residues	Gaseous Residues
Premanufacture	1,1	1,2	1,3	1,4	1,5
Product Manufacture	2,1	2,2	2,3	2,4	2,5
Product Delivery	3,1	3,2	3,3	3,4	3,5
Product Use	4,1	4,2	4,3	4,4	4,5
Refurbishment, Recycling, Disposal	5,1	5,2	5,3	5,4	5,5

The numbers in each box are the matrix element indices.

longer satisfactory because of obsolescence, component degradation, or changed business or personal decisions is refurbished or discarded.

   An assessment system that we recommend for its efficiency and demonstrated usefulness has as its central feature a 5x5 assessment matrix, the Environmentally Responsible Product Assessment Matrix, one dimension of which is life-cycle stage and the other of which is environmental concern (Table 12.3). In use, the Design for Environment assessor studies the product design, manufacture, packaging, in-use environment, and likely disposal scenario and assigns to each element of the matrix an integer rating from 0 (highest impact, a very negative evaluation) to 4 (lowest impact, an exemplary evaluation). In essence, what the assessor is doing is providing a figure of merit to represent the estimated result of the more formal life cycle assessment inventory analysis and impact analysis stages. She or he is guided in this task by experience, a design and manufacturing survey, appropriate checklists, and other information (*Appendix A*). The process described here is purposely qualitative and utilitarian, but does provide a numerical end point against which to measure improvement.

   Once an evaluation has been made for each matrix element, the overall Environmentally Responsible Product Rating (R_{ERP}) is computed as the sum of the matrix element values:

$$R_{ERP} = \sum_i \sum_j M_{i,j} \qquad (12.1)$$

Since there are 25 matrix elements, a maximum product rating is 100.

12.4 ASSESSING GENERIC AUTOMOBILES OF YESTERDAY AND TODAY

Automobiles have environmental impacts during all their life stages, in contrast to many other products such as furniture or roofing materials. The greatest impacts result from the combustion of gasoline and the release of tailpipe emissions during the driving cycle. However, there are other aspects of the product that affect the environment, such as the dissipative use of oil and other lubricants, the discarding of tires and other spent parts, and the ultimate retirement of the vehicle.

As a demonstration of the operation of the tools described above, we perform environmentally-responsible product assessments on generic automobiles of the 1950s and 1990s. Some of the relevant characteristics of the vehicles were given in Table 7.1.

Premanufacturing, the first life stage, treats impacts on the environment as a consequence of the actions needed to extract materials from their natural reservoirs, transport them to processing facilities, purify or separate them by such operations as ore smelting and petroleum refining, and transport them to the manufacturing facility. Where components are sourced from outside suppliers, this life stage also incorporates assessment of the impacts arising from component manufacture. The ratings that we assign to this life stage of generic vehicles from each epoch are given below, where the two numbers in parentheses refer to the matrix element indices shown in Table 12.3. The higher (that is, more favorable) ratings for the 1990 vehicle are mainly due to improvements in the environmental aspects of mining and smelting technologies, improved efficiency of the equipment and machinery used, and the increased use of recycled material.

Premanufacture Ratings

Element Designation	Element Value and Explanation
1950s auto: Matls. choice (1,1)	2 (Few toxics are used, but most materials are virgin)
Energy use (1,2)	2 (Virgin material shipping is energy-intensive)
Solid residue (1,3)	3 (Iron and copper ore mining generates substantial solid waste)
Liq. residue (1,4)	3 (Resource extraction generates moderate amounts of liquid waste)
Gas residue (1,5)	2 (Ore smelting generates significant amounts of gaseous waste)
1990s auto: Matls. choice (1,1)	3 (Few toxics are used; much recycled material is used)
Energy use (1,2)	3 (Virgin material shipping is energy-intensive)
Solid residue (1,3)	3 (Metal mining generates solid waste)
Liq. residue (1,4)	3 (Resource extraction generates moderate amounts of liquid waste)
Gas residue (1,5)	3 (Ore processing generates moderate amounts of gaseous waste)

The second life stage is product manufacture. The basic automotive manufacturing process has changed little over the years but much has been done to improve its environmental responsibility. One potentially high-impact area is the paint shop, where various chemicals are used to clean the parts and volatile organic emissions are generated during the painting process. There is now greater emphasis on treatment and recovery of waste water from the paint shop and the switch from low-solids to high-solids paint has done much to reduce the amount of material emitted. With respect to material fabrication there is currently better utilization of material (partially due to better analytical techniques for designing component parts) and a greater emphasis on reusing scraps and trimmings from the various fabrication processes. Finally, the productivity of the entire manufacturing process has been improved, substantially less energy and time being required to produce each automobile.

Product Manufacture Ratings

Element Designation	Element Value and Explanation
1950s auto: Matls. choice (2,1)	0 (CFCs used for metal parts cleaning)
Energy use (2,2)	1 (Energy use during manufacture is high)
Solid residue (2,3)	2 (Lots of metal scrap and packaging scrap produced)
Liq. residue (2,4)	2 (Substantial liquid residues from cleaning and painting)
Gas residue (2,5)	1 (Volatile hydrocarbons emitted from paint shop)
1990s auto: Matls. choice (2,1)	3 (Good materials choices, except for lead solder waste)
Energy use (2,2)	2 (Energy use during manufacture is fairly high)
Solid residue (2,3)	3 (Some metal scrap and packaging scrap produced)
Liq. residue (2,4)	3 (Some liquid residues from cleaning and painting)
Gas residue (2,5)	3 (Small amounts of volatile hydrocarbons emitted)

The environmental concerns at the third life stage, product packaging and transport, include the manufacture of the packaging material, its transport to the manufacturing facility, residues generated during the packaging process, transportation of the finished and packaged product to the customer, and (where applicable) product installation. This aspect of the automobile's life cycle is benign relative to the vast majority of products sold today, since automobiles are delivered with negligible packaging material. Nonetheless, some environmental burden is associated with the transport of a large, heavy product. The

Product Delivery Ratings

Element Designation	Element Value and Explanation
1950s auto: Matls. choice (3,1)	3 (Sparse, recyclable materials used during packaging and shipping)
Energy use (3,2)	2 (Over-the-road truck shipping is energy-intensive)
Solid residue (3,3)	3 (Small amounts of packaging during shipment could be further minimi
Liq. residue (3,4)	4 (Negligible amounts of liquids are generated by packaging and shippin
Gas residue (3,5)	2 (Substantial fluxes of greenhouse gases are produced during shipment)
1990s auto: Matls. choice (3,1)	3 (Sparse, recyclable materials used during packaging and shipping)
Energy use (3,2)	3 (Long-distance land and sea shipping is energy-intensive)
Solid residue (3,3)	3 (Small amounts of packaging during shipment could be further minimi
Liq. residue (3,4)	4 (Negligible amounts of liquids are generated by packaging and shippin
Gas residue (3,5)	3 (Moderate fluxes of greenhouse gases are produced during shipment)

slightly higher rating for the 1990s automobile is due mainly to the better design of auto carriers (more vehicles per load) and the increase in fuel efficiency of the transporters.

The fourth life stage, product use, includes impacts from consumables (if any) or maintenance materials (if any) that are expended during customer use. Significant progress has been made in automobile efficiency and reliability, but automotive use continues to have a very high negative impact on the environment. The increase in fuel efficiency and more effective conditioning of exhaust gases accounts for the 1990s automobile achieving higher ratings, but clearly there is still room for improvement.

Customer Use Ratings

Element Designation	Element Value and Explanation
1950s auto: Matls. choice (4,1)	1 (Petroleum is a resource in limited supply)
Energy use (4,2)	0 (Fossil fuel energy use is very large)
Solid residue (4,3)	1 (Significant residues of tires, defective or obsolete parts)
Liq. residue (4,4)	1 (Fluid systems are very leaky)
Gas residue (4,5)	0 (No exhaust gas scrubbing; high emissions)
1990s auto: Matls. choice (4,1)	1 (Petroleum is a resource in limited supply)
Energy use (4,2)	2 (Fossil fuel energy use is large)
Solid residue (4,3)	2 (Modest residues of tires, defective or obsolete parts)
Liq. residue (4,4)	3 (Fluid systems are somewhat dissipative)
Gas residue (4,5)	2 (CO_2, lead [in some locales])

The fifth life stage assessment includes impacts during product refurbishment and as a consequence of the eventual discarding of modules or components deemed impossible or too costly to recycle. Most modern automobiles are recycled (some 95% currently enter the recycling system), and from these approximately 75% by weight is recovered for used parts or returned to the secondary metals market. There is a viable used parts market and most cars are stripped of reusable parts before they are discarded. Improvements in recovery technology have made it easier and more profitable to separate the automobile into its component materials.

In contrast to the 1950s, at least two aspects of modern automobile design and construction are retrogressive from the standpoint of their environmental implications. One is the increased diversity of materials used, mainly the increased use of plastics. The second aspect is the increased use of welding in the manufacturing process. In the vehicles of the 1950s, a body-on-frame construction was used. This approach was later switched to a unibody construction technique in which the body panels are integrated with the chassis. Unibody construction requires about four times as much welding as does body-on-frame construction, plus substantially increased use of adhesives. The result is a vehicle that is stronger, safer, and uses less structural material, but is much less easy to disassemble.

Refurbishment/Recycling/Disposal Ratings

Element Designation   Element Value and Explanation

1950s auto: Matls. choice (5,1)   3 (Most materials used are recyclable)
      Energy use (5,2)   2 (Moderate energy use required to disassemble and recycle materials)
      Solid residue (5,3)   2 (A number of components are difficult to recycle)
      Liq. residue (5,4)   3 (Liquid residues from recycling are minimal)
      Gas residue (5,5)   1 (Recycling commonly involves open burning of residues)

1990s auto: Matls. choice (5,1)   3 (Most materials recyclable, but sodium azide presents difficulty)
      Energy use (5,2)   2 (Moderate energy use required to disassemble and recycle materials)
      Solid residue (5,3)   3 (Some components are difficult to recycle)
      Liq. residue (5,4)   3 (Liquid residues from recycling are minimal)
      Gas residue (5,5)   2 (Recycling involves some open burning of residues)

The completed matrices for the generic 1950s and 1990s automobile are illustrated in Table 12.4. Examine first the values for the 1950s vehicle so far as life stages are concerned. The column at the far right of the table shows moderate environmental stewardship during resource extraction, packaging and shipping, and refurbishment/recycling/disposal. The ratings during manufacturing are poor, and during customer use are abysmal. The overall rating of 46 is far below what might be desired. In contrast, the overall rating for the 1990s vehicle is 68, much better than that of the earlier vehicle but still leaving plenty of room for improvement.

The matrix displays provide a useful overall assessment of a design, but a more succinct display of DFE design attributes is provided by the "target plots" shown in Fig. 12.7. To construct the plots, the value of each element of the matrix is plotted at a specific angle. (For a 25–element matrix, the angle spacing is $360/25 = 14.4°$.) A good product or process shows up as a series of dots bunched in the center, as would occur on a rifle target in which each shot was aimed accurately. The plot makes it easy to single out points far removed from the bulls-eye and to mark their topics out for special attention. Furthermore, the comparison of target plots for alternative designs of the same product permits quick comparisons of environmental responsibility. The product and process design teams can then select among design options, and can consult the checklists and protocols for information on improving individual matrix element ratings.

12.5 THE WEIGHTED MATRIX

Much useful information is provided by the unweighted matrix and target plot, yet there is an underlying uneasiness in that equal importance has been given to, for example, the "packaging and transport" and "in-service" life stages. Since the latter will certainly have a greater environmental effect during the life of the 1950s automobile, it is appropriate to contemplate some form of matrix element weighting to reflect this information. Thus, instead of the initial approach, in which each matrix element had a maximum value of 4 and the overall assessment value was given by summing those impact assessments, a set of

Table 12.4 Environmentally Responsible Product Assessments for the Generic 1950s and 1990s Automobiles*

Life Stage	Environmental Concern					
	Materials Choice	Energy Use	Solid Residues	Liquid Residues	Gaseous Residues	Total
Resource Extraction	2	2	3	3	2	12/20
	3	3	3	3	3	15/20
Product Manufacture	0	1	2	2	1	6/20
	3	2	3	3	3	14/20
Product Delivery	3	2	3	4	2	14/20
	3	3	3	4	3	16/20
Product Use	1	0	1	1	0	3/20
	1	2	2	3	2	10/20
Refurbishment, Recycling, Disposal	3	2	2	3	1	11/20
	3	2	3	3	2	13/20
Total	9/20	7/20	11/20	13/20	6/20	46/100
	13/20	12/20	14/20	16/20	13/20	68/100

* Upper numbers refer to the 1950s automobile, lower numbers to the 1990s automobile.

weighting factors $\omega_{i,j}$ is sought that reflects differences in life stage impact such that each matrix element will have a value given by $\omega_{i,j}M_{i,j}$ and for which the overall rating is given by

$$R_{ERP} = \sum_i \sum_j \omega_{i,j} M_{i,j} \tag{12.2}$$

Determining in a comprehensive way the appropriate life stage weightings requires a complete life-cycle assessment, and is in conflict with the desire to streamline the LCA process. The assessment can be intuitively improved, however, by choosing the life stage likely to produce the most severe environmental impacts and arbitrarily weighting that life stage as one half the total assessment value, with the other four life stages arbitrarily weighted at one-eighth the total value. The weighting factors are then those shown in Table 12.5.

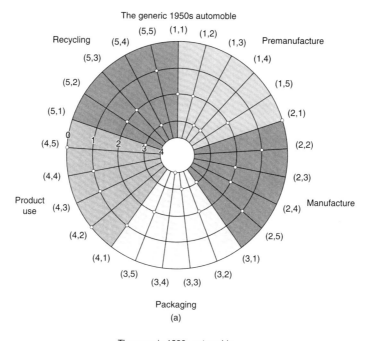

The generic 1950s automoble

(a)

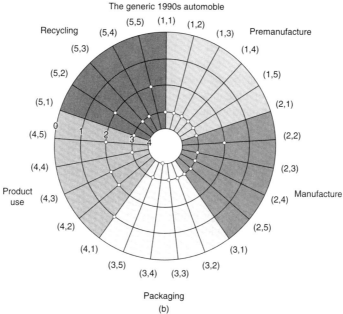

The generic 1990s automoble

(b)

Figure 12.7   Comparitive target plots for the display of the environmental impacts of the generic automobile of the 1950s and of the 1990s.

Table 12.5 $\omega_{i,j}$ Values for the Singly-Weighted Matrix

Life Stage	Environmental Concern				
	Materials Choice	Energy Use	Solid Residues	Liquid Residues	Gaseous Residues
Premanufacture	0.625	0.625	0.625	0.625	0.625
Product Manufacture	0.625	0.625	0.625	0.625	0.625
Product Delivery	0.625	0.625	0.625	0.625	0.625
Product Use	2.5	2.5	2.5	2.5	2.5
Refurbishment, Recycling, Disposal	0.625	0.625	0.625	0.625	0.625

If the matrix element values are recomputed using the weighting factors, the resultant evaluation for the generic 1950s automobile is that shown in Table 12.6, where the increase in influence of life stage 4 is evident and where the overall product rating has dropped from 46 to 33. This result can be made even more apparent by the target plot. However, the construction of the target plot now requires an additional step in order to demonstrate the degree of departure from optimum of a matrix element value. The procedure is to compute the "deficit matrix", the difference between a perfect score and the actual score, given for each matrix element by

$$\delta_{i,j} = [\omega_{i,j}M_{i,j}]_{max} - \omega_{i,j}M_{i,j} \qquad (12.3)$$

In the 1950s automobile example, the resulting target plot is shown in Fig. 12.8, where, by comparison with Fig. 12.7, the deficiencies in environmental responsibility during the in-service life stage are quite obvious.

12.6 THE DOUBLY-WEIGHTED MATRIX

Just as some life stages produce larger environmental impacts than others, some environmental impacts are of more concern than others. The EPA Science Advisory Board has suggested the following as ranking parameters to sort environmental impacts into priority order:

Table 12.6   The Singly-Weighted Environmentally Responsible Product
Assessment for the Generic 1950s Automobile

Life Stage	Environmental Concern					
	Materials Choice	Energy Use	Solid Residues	Liquid Residues	Gaseous Residues	Total
Premanufacture	1.25	1.25	1.25	1.25	1.25	6.3/12.5
Product Manufacture	0	0.63	1.25	1.25	0.63	3.8/12.5
Product Delivery	1.88	1.25	1.88	2.50	1.25	8.8/12.5
Product Use	2.50	0	2.50	2.50	0	7.50/50
Refurbishment, Recycling, Disposal	1.88	1.25	1.25	1.88	0.63	6.8/12.5
Total	7.5/20	4.4/20	8.1/20	9.4/20	3.8/20	33/100

- The spatial scale of the impact (large scales being worse than small).
- The severity of the hazard (more highly toxic substances being of more concern than less highly toxic substances).
- The degree of exposure (well sequestered substances being of less concern than readily mobilized substances).
- The penalty for being wrong (longer remediation times being of more concern than shorter times).

For a specific product, the priority environmental concern is thus some combination of the magnitude of the product's impacts and the degree to which the impacts satisfy the high-risk criteria. Determining priorities rigorously requires, as before, a comprehensive and defendable life cycle assessment.

In the case of the 1950s automobile, one could make a case for either energy or gaseous emissions being the dominant impact, but for demonstration purposes energy will be selected here. As with the life stage weightings, the energy use impact is arbitrarily assigned half the total assessment value, while the other environmental concerns receive one-eighth each. The weighting factors $\phi_{i,j}$ are then as shown in Table 12.7, and the composite weighting factors $\Omega_{i,j}$ (Table 12.8) computed as

$$\Omega_{i,j} = \omega_{i,j}\phi_{i,j} \tag{12.4}$$

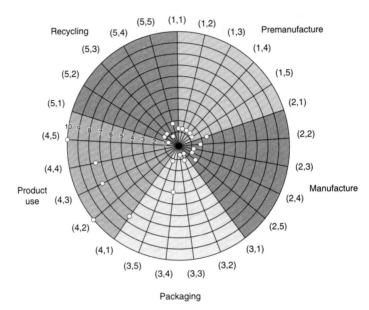

Figure 12.8   A target plot for the display of the weighted matrix assessment results for the generic automobile of the 1950s.

The doubly-weighted matrix element values are then given by the product $\omega_{i,j}\phi_{i,j}M_{i,j}$ and the overall rating by

$$R_{ERP} = \sum_i \sum_j \Omega_{i,j}M_{i,j} \qquad (12.5)$$

The matrix result for the 1950s automobile is shown in Table 12.9, the overall rating now having dropped from 33 to 29 and the importance of energy use now being emphasized.

The target plot for the doubly-weighted matrix is formed in a manner similar to that for the singly-weighted matrix, deficits being computed as

$$\Delta_{i,j} = [\Omega_{i,j}M_{i,j}]_{max} - \Omega_{i,j}M_{i,j} \qquad (12.6)$$

The resulting target plot in Fig. 12.9 shows dramatically the importance of the in-service life stage, the energy use environmental concern, and (most importantly) the failure of the 1950s vehicle to reflect these priorities. This failure is, of course, unsurprising, as such vehicles were designed before environmental concerns were widespread, and when gasoline was considered an unlimited resource.

Table 12.7 $\phi_{i,j}$ Values for the Doubly-Weighted Matrix

Life Stage	Environmental Concern				
	Materials Choice	Energy Use	Solid Residues	Liquid Residues	Gaseous Residues
Premanufacture	0.625	2.5	0.625	0.625	0.625
Product Manufacture	0.625	2.5	0.625	0.625	0.625
Product Delivery	0.625	2.5	0.625	0.625	0.625
Product Use	0.625	2.5	0.625	0.625	0.625
Refurbishment, Recycling, Disposal	0.625	2.5	0.625	0.625	0.625

Table 12.8 $\Omega_{i,j}$ Values for the Doubly-Weighted Matrix

Life Stage	Environmental Concern				
	Materials Choice	Energy Use	Solid Residues	Liquid Residues	Gaseous Residues
Premanufacture	0.39	1.56	0.39	0.39	0.39
Product Manufacture	0.39	1.56	0.39	0.39	0.39
Product Delivery	0.39	1.56	0.39	0.39	0.39
Product Use	1.56	6.25	1.56	1.56	1.56
Refurbishment, Recycling, Disposal	0.39	1.56	0.39	0.39	0.39

Table 12.9 The Doubly-Weighted Environmentally Responsible Product Assessment for the Generic 1950s Automobile

Life Stage	Environmental Concern					Total
	Materials Choice	Energy Use	Solid Residues	Liquid Residues	Gaseous Residues	
Premanufacture	0.78	3.13	0.78	0.78	0.78	6.25/12.5
Product Manufacture	0	1.56	0.78	0.78	0.39	3.51/12.5
Product Delivery	1.17	3.13	1.17	4	0.78	7.81/12.5
Product Use	1.56	0	1.56	1.56	0	4.68/50
Refurbishment, Recycling, Disposal	1.17	3.13	0.78	1.17	0.39	6.64/12.5
Total	4.68/12.5	10.95/50	5.07/12.5	5.85/12.5	2.34/12.5	28.89/100

12.7 ASSESSING THE ROADWAY INFRASTRUCTURE OF YESTERDAY AND TODAY

As with the automobile, the infrastructure has environmental impacts during all its life stages, and these can be evaluated also by the streamlined assessment tools described above. In harmony with the automobile assessment, we perform here environmentally-responsible systems assessments on a unit length of high-speed generic roadway infrastructure of the 1950s and 1990s. (The use of the term "roadway" here is to restrict the analysis by omitting fuel stations, repair shops, and so forth.) The magnitude of the unit length is not critical, since we will consider only qualitative aspects, but one can think in terms of 5–10 km as a unit length that would encompass all relevant aspects of the roadway and its related components. Some of the characteristics of our generic infrastructure systems are given in Table 12.10. In overview, the 1950s infrastructure was a multi-lane highway representing a transition of an older two-lane roadway into a higher-speed roadway. The necessary road widening was unconstrained by environmental considerations, as was the subsequent asphalt paving and ancillary roadway improvements such as signs and occasional guard rails. Patching and other surface repair was performed as needed.

The 1990s infrastructure, in contrast, is a limited-access highway of the type represented by the German Autobahn, the Italian Autostrada, or the US Interstate Highway. The

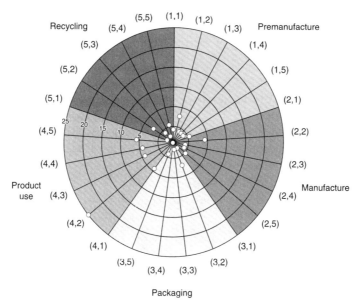

Figure 12.9   A target plot for the display of the doubly-weighted matrix assessment results for the generic auto-mobile of the 1950s.

surface is made of reinforced concrete, as are the bridges and overpasses, and extensive signage and overhead lighting is used at roadway exits. Sand and salt are spread extentively on the roadway in the winter months to improve traction. Every several years the roadway surface is removed and new roadway is laid down.

A significant factor in the degree of environmental responsibility of an infrastructure system is the sites selected and the way in which those sites are developed. A large fraction of the transportation infrastructure is constrained to be located in or near urban areas. For infrastructure of any kind built on land previously undeveloped as industrial or commercial sites, ecological impacts on regional biodiversity can be anticipated, as well as added air emissions from new transportation and utility infrastructures. These effects can be minimized with attention to working with existing infrastructures and developing the site with the maximum area being left in natural form. Nonetheless, given the requirement for infrastructure to follow need, there is a strong societal component in how that infrastructure is developed. This site selection and development process is the first infrastructure life stage. The ratings that we assign to this life stage of generic infrastructure from each epoch are given below, where the two numbers in parentheses refer to the matrix element indices shown in Table 12.11. The higher (that is, more favorable) ratings for the 1990 infrastructure are mainly due to improvements in the environmental aspects of site development.

Table 12.10   Characteristics of Automotive Infrastructure
of the 1950s and 1990s

Characteristic	ca. 1950s	ca. 1990s
Roadway Characteristics		
Type	Unlimited access	Limited access
Speed limit	90 km/hr	110 km/hr
Surface	Asphalt	Reinf. concrete
Route	Historic road	Farmland, wetland
Site Selection and Development		
Location	Unconstrained	Moderate control
Solid residues	No control	Minor control
Energy use	Very high	Very high
Infrastructure Manufacture		
Stone	Heavy use	Heavy use
Minerals	Light use	Heavy use
Steel	Light use	Heavy use
Petroleum	Heavy use	Moderate use
Energy use	Very high	Very high
Infrastructure Use		
Energy use, patching	High	High
Energy use, maintenance	Low	Moderate
Solid residues, patching	No control	Minor control
Painting	Pb, high VOC	Pb, moderate VOC
Road sand, salt	Minor amount	Large amount
Comp. Infrastructure Impacts		
Parking facilities	Moderate impact	High impact
Secondary roads	Moderate impact	High impact
Sewer facilities	Minor impact	High impact
Local hydrology	Moderate impact	High impact
Infrastructure Recycling		
Asphalt	None	Modest amount
Concrete	None	Minor amount
Steel	Small amount	Substantial amount

Table 12.11   The Environmentally-Responsible Infrastructure System Matrix

Facility activity	Environmental concern				
	Ecological impacts	Energy use	Solid residues	Liquid residues	Gaseous residues
Site Selection and development	1,1	1,2	1,3	1,4	1,5
Infrastructure manufacture	2,1	2,2	2,3	2,4	2,5
Infrastructure use	3,1	3,2	3,3	3,4	3,5
Complementary systems	4,1	4,2	4,3	4,4	4,5
Infrastructure end of life	5,1	5,2	5,3	5,4	5,5

The numbers in each box are the matrix element indices.

Site Selection and Development Ratings

Element Designation	Element Value and Explanation
1950s: Biodiversity/materials (1,1)	0 (No concern for site effects)
Energy use (1,2)	1 (Very high energy consumption)
Solid residue (1,3)	0 (No solid residue control)
Liq. residue (1,4)	1 (Minimal liquid residue control)
Gas residue (1,5)	1 (Minimal gaseous residue control)
1990s: Biodiversity/materials (1,1)	1 (Large land areas altered)
Energy use (1,2)	1 (Very high energy consumption)
Solid residue (1,3)	1 (Limited solid residue control)
Liq. residue (1,4)	2 (Minimal liquid residue control)
Gas residue (1,5)	2 (Minimal gaseous residue control)

The second life stage is infrastructure manufacture. Many of the basic automotive infrastructure manufacturing processes have changed little over the years, although modest efforts have been made to improve their environmental responsibility. The big change is the progression from unlimited access asphalt surfaces to limited-access highways constructed largely of reinforced concrete and with overpasses, bridges, and entrance and exit ramps of the same material. Such a roadway requires more materials throughout; overall energy use is higher than for an asphalt surface, but the rates of gaseous emissions during construction are lower.

Where infrastructure is manufactured on-site, as with asphalt or concrete, the concerns encompass extraction, processing, and transport of the materials delivered to the site and used in the manufacture of pavements or structures. Where infrastructure materials or components are manufactured elsewhere and consumed or installed on site, as with cement, bridges, girders, light stanchions, or signage, the concerns encompass manufacture of the material, packaging, transportation of the product to the customer, and product installation.

In the case of the 1950s roadway, the refinery manufacture of bitumen was energy-intensive and greenhouse gas emissions high. Lights, signage, and the like were not used to a great extent, and were not overpackaged, so site residues from shipping and installation were modest.

The 1990s roadway of concrete required large volumes of cement, the manufacture of which is very energy intensive and which emits very large quantities of greenhouse gases. Infrastructure components manufactured off-site are abundant, and their packaging, shipping, and installation create significant impacts.

Infrastructure Manufacture Ratings

Element Designation	Element Value and Explanation
1950s: Biodiversity/materials (2,1)	3 (Most materials are relatively benign)
Energy use (2,2)	2 (High energy consumption)
Solid residue (2,3)	2 (Substantial solid residues from materials processing)
Liq. residue (2,4)	2 (Substantial liquid residues from materials processing)
Gas residue (2,5)	2 (Moderate gaseous residues from asphalt manufacture)
1990s: Biodiversity/materials (2,1)	2 (Most materials relatively benign, but used prolifically)
Energy use (2,2)	1 (Very high energy consumption)
Solid residue (2,3)	2 (Substantial solid residues from materials processing)
Liq. residue (2,4)	2 (Substantial liquid residues from materials processing)
Gas residue (2,5)	3 (Modest gaseous residues from construction vehicles)

Life stage 3a comprises impacts from consumables or maintenance materials that are expended while keeping the automotive infrastructure well-maintained and functioning properly. The two infrastructures differ dramatically. The 1950s asphalt surface required modest amounts of virgin materials to renew it. (Unlike today, asphalt was not then recycled.) Small amounts of sand and salt were used to maintain winter traction. The exhausts on the maintenance vehicles were uncontrolled.

The 1990s roadway is more difficult to maintain once it is degraded. The broken pavement or rusted reinforcing steel cannot be horizontally recycled, so new material is required. Road sand and salt are used extensively. Some emission control is in effect for maintenance vehicles.

Infrastructure Use Ratings

Element Designation	Element Value and Explanation
1950s: Biodiversity/materials (3,1)	2 (Minor road salting, little matls. reuse)
Energy use (3,2)	2 (Fossil fuel energy use in maintenance is large)
Solid residue (3,3)	3 (Sand, degraded pavement are residues)
Liq. residue (3,4)	3 (Modest fluid emissions)
Gas residue (3,5)	1 (High maintenance vehicle exhaust emissions)
1990s: Biodiversity/materials (3,1)	0 (Extensive road salting, no matls. reuse)
Energy use (3,2)	1 (Fossil fuel energy use in maintenance is large)
Solid residue (3,3)	1 (Sand, salt, degraded pavement, steel)
Liq. residue (3,4)	3 (Modest fluid emissions)
Gas residue (3,5)	2 (Substantial maintenance vehicle exhaust emissions)

Both the 1950s and 1990s roadway infrastructures have significant interactions with other infrastructure components, the principal distinction being in the scale and form of manufacture and maintenance in the different epochs. The impacts of most concern are on related transportation components, especially secondary roads and parking facilities. Also of significance are impacts related to water runoff, both the storm sewers that must be constructed to handle the water, especially in the case of the limited access roadway, and the adversed impacts of pulsed water flow to local ecosystems. Secondary considerations include modifications of auto-related components: petrol and bitumen distribution systems, auto repair facilities, and the like.

Complementary Infrastructure Ratings

Element Designation	Element Value and Explanation
1950s: Biodiversity/materials (4,1)	2 (Moderate hydrologic impacts)
Energy use (4,2)	3 (Exacerbates energy consumption)
Solid residue (4,3)	2 (Local road and parking expansion)
Liq. residue (4,4)	3 (Minor sewerage impacts)
Gas residue (4,5)	1 (Local road and parking expansion)
1990s: Biodiversity/materials (4,1)	1 (Major hydrologic impacts)
Energy use (4,2)	2 (High energy consumption encouraged)
Solid residue (4,3)	1 (Local road and parking expansion)
Liq. residue (4,4)	1 (Major sewerage impacts)
Gas residue (4,5)	0 (Local road and parking expansion)

The final life stage assessment includes impacts during infrastructure refurbishment and as a consequence of the eventual discarding of components or entire systems deemed obsolete and too costly to recycle. Neither the 1950s roadway nor that of the 1990s is specifically designed with end of life in mind, but the asphalt surface is horizontally recyclable to other roadways or to itself; concrete is not. The reinforcing steel in the 1990s roadway is difficult to separate from the concrete matrix, tends to be corroded by road salt over time, and thus has little value. Signage, lighting stanchions, and other steel components can generally be recycled.

Refurbishment/Recycling/Disposal Ratings

Element Designation	Element Value and Explanation
1950s: Biodiversity/materials (5,1)	3 (Most materials used are recyclable)
Energy use (5,2)	2 (Substantial energy use required to recycle materials)
Solid residue (5,3)	3 (Some materials and components discarded)
Liq. residue (5,4)	4 (Liquid residues from disposal are minimal)
Gas residue (5,5)	3 (Gaseous residues from recycling or disposal are moderate)
1990s: Biodiversity/materials (5,1)	2 (Materials used are difficult to recycle)
Energy use (5,2)	2 (Substantial energy use required to recycle materials)
Solid residue (5,3)	2 (Many materials and components discarded)
Liq. residue (5,4)	4 (Liquid residues from final life stage are minimal)
Gas residue (5,5)	3 (Gaseous residues from final life stage are moderate)

The completed matrices for the generic 1950s and 1990s automotive infrastructure are illustrated in Table 12.12. Examine first the values for the 1950s infrastructure so far as life stages are concerned. The column at the far right of the table shows moderate environmental stewardship during most life stages. The ratings during site selection and preparation are very poor; no life stages demonstrate truly responsible performance. The overall rating of 51 is far below what might be desired. In the case of the 1990s infrastructure, roadway manufacture and end of life are only passable, and the site selection and preparation, infrastructure use, and complementary infrastructure component impact life stages show poor environmental performance. The overall rating is 42, even lower than that of the earlier epoch. The principal reasons for these low ratings are the high degree of ecosystem disturbance during site preparation, severe environmental impacts during infrastructure use, significant adverse impacts on complementary infrastructure components, and almost complete failure to consider optimizing the end of life of the roadways in developing the materials properties and infrastructure designs.

The matrix displays provide a useful overall assessment of a design, but a more succinct display of DFE design attributes is provided by the "target plots" shown in Fig. 12.10. In both cases, the target plots show opportunities for improvement at all life stages and over all environmental concerns.

12.8 SUMMARY

Unlike Chapters 7–11, which offered advice for improving specific aspects of the environmental impacts of the automobile, this chapter evaluates the entire vehicle and the entire infrastructure by quantitative and semi-quantitative techniques. The detailed EPS method is beneficial but intricate, requiring lengthy and sometimes contentious efforts by scientists

Table 12.12   Environmentally Responsible Assessments
for the Generic 1950s and 1990s Automotive Infrastruture*

Life Stage	Environmental Concern					
	Biodiversity/ Materials	Energy Use	Solid Residues	Liquid Residues	Gaseous Residues	Total
Site selection and Preparation	0 1	1 1	0 1	1 2	1 2	3/20 7/20
Infrastructure manufacture	3 2	2 1	2 2	2 2	2 3	11/20 10/20
Infrastructure use	2 0	2 1	3 1	3 3	1 2	11/20 7/20
Complementary systems	2 1	3 2	2 1	3 1	1 0	11/20 5/20
Refurbishment, Recycling, Disposal	3 2	2 2	3 2	4 4	3 3	15/20 13/20
Total	10/20 6/20	10/20 7/20	10/20 7/20	13/20 12/20	8/20 10/20	51/100 42/100

* Upper numbers refer to the 1950s infrastructure, lower numbers to the 1990s infrastructure.

and engineers. The underlying data require continuing upgrading, but, once computerized, the system is relatively easy to access by design teams. The contrasting matrix technique is efficient and easily incorporated into corporate operations. Although some aspects of it require professional judgment, most of the vital environmental concerns are addressed and highlighted for the design team. While it is apparent that the techniques themselves will benefit from further development, even in their present state both are useful tools with which to guide the design engineer.

The contrast between the environmental attributes of simple and complex products is nowhere more apparent than in assessing their impacts over their entire life cycles. In general, the principal environmental impacts of simple products (food, shampoo, concrete road surfaces) tend to result from the choice of materials from which they are manufactured. In the case of complex products(automobiles, electronic highway signage), the principal impacts tend to be related to the level of overall environmental expertise of the design teams and the role the product plays in the economy. The automobile is not an environmental issue because of its 1500 kilograms of materials, but in the impacts related to the functioning of those materials; a completely different perspective applies to, for example, 1500 kilograms of brown paper grocery bags.

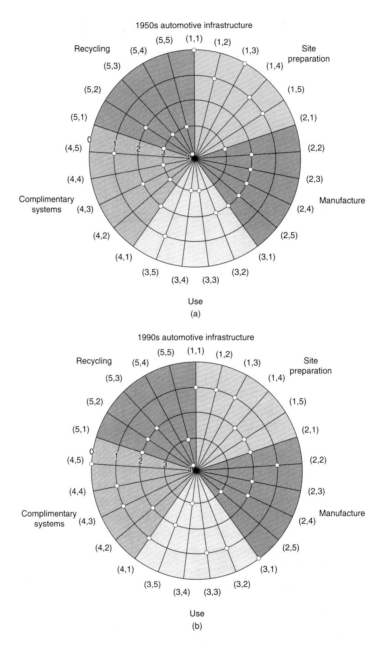

Figure 12.10   Comparitive target plots for the display of the environmental impacts of the generic automotive infrastructure of the 1950s and of the 1990s.

The results of the LCA and SLCA exercises show that good progress is being made in automotive design insofar as environmental attributes are concerned. In particular, the substantial gains between 1950 and 1990 are obvious. Nonetheless, there is room for much further progress. Tomorrow's automotive engineers face equally great design for environment challenges as compared with those faced by the engineers of yesterday and today.

The situation with infrastructure is still more interesting. Modern high-speed highways are undeniably successful in moving more vehicles than did the roads of a half-century ago, and doing so at significantly higher average speeds. However, from an environmental standpoint the modern highway is not only not exemplary, but is even worse than yesterday's poorly-performing version. There is no doubt that the highway engineer faces a difficult balancing act in optimizing the sometimes conflicting demands of cost, roadway performance, and pressures for accomplishing improvements within an existing infrastructure rather than starting from scratch. Nonetheless, almost no effort has been given to the development of alternative roadway construction materials, recycling improvements, and transformations of in-use aspects. It is interesting to speculate on how much improvement will come from better civil engineering and how much from enhanced information systems, which have the effect of creating a more complex product. An existing highway is a relatively simple product as we have defined the term. An intelligent highway system, however, begins to look more like a complex product. Much remains to be done to make the roadway infrastructure more environmentally responsible.

SUGGESTED READING

Environmental Protection Agency Science Advisory Board, *Reducing Risk: Setting Priorities and Strategies for Environmental Protection*, Report SAB-EC-90–021 and 021A, Washington, D.C., 1990.

Graedel, T.E., and B.R. Allenby, *Design for Environment*, Upper Saddle River, N.J.: Prentice-Hall, Inc., 175 pp., 1996.

Graedel, T.E.; B.R. Allenby, P.R. Comrie, Matrix approaches to abridged life cycle assessment, *Environmental Science and Technology, 20*, 134A-139A, 1995.

Society of Environmental Toxicology and Chemistry (SETAC), *A Technical Framework for Life-Cycle Assessment*, Washington, D. C., 1991.

Vigon, B.W., D.A. Tolle, B.W. Cornaby, H.C. Latham, C.L. Harrison, T.L. Boguski, R.G. Hunt, and J.D. Sellers, *Life-Cycle Assessment: Inventory Guidelines and Principles*, EPA/600/R-92/036, Cincinnati, OH: U.S. Environmental Protection Agency, 1992.

EXERCISES

12.1 Consider the environmental impacts currently associated with automobile use, including all emissions and directly related activity in the construction and petroleum sectors.

  a. Develop your own set of criteria by which to prioritize environmental impacts. Explain and defend your choices.

b. Using your criteria, rank the environmental impacts you have identified into three categories: major, important, and minor.

c. Determine which of the environmental impacts you have identified as major can be substantially mitigated using current technologies, and which require significant cultural changes before they can be successfully addressed.

d. Using the results of your analysis, develop a research agenda to reduce the environmental impacts associated with the automobile culture. Estimate the resources required to carry out each suggested research program, and the benefits which may be anticipated from its successful completion.

12.2 (a) Suppose that global warming is thought less likely to occur than had previously been assumed, and that as a result the ELU/kg for product use is lowered to 0.6. What effect does this have on the comparative ratings of the two front ends of Fig. 12.4? (b) A new high-strength honeycomb steel has been developed and is being considered for use in automobile front ends. Rather than the steel front end weighing 6.0 kg, a satisfactory front end weighing only 4.0 kg can be formed from 6.0 kg of the new steel. Since the new steel's improved properties, which are due to added trace alloying elements, have negligible effects on processing or recycling, the same ELU/kg assessments apply to those stages (Table 12.2). Compute the ELU values for the new front end and compare them with the two options in the table. (c) Assume that the global warming revision of part (a) occurs as well as the availability of the new honeycomb steel front end. What effect do these two changes taken together have on the relative impact results? (d) Suppose that the shortage of petroleum (used as a feedstock for the manufacture of plastic composites) became so great that the ELU/kg of the composite materials was set to 1.90 and that the honeycomb steel front end was available. What effect do these two changes taken together have on the relative impact results? (e) What are the messages to designers implied in the analyses in the earlier parts of this exercise?

12.3 For the doubly-weighted matrix, the life stage and environmental concern that were selected were weighted so as to comprise half the total, all other life stages and concerns comprising the other half. If the chosen factors were weighted so as to comprise one-third the total rather than one-half, compute the weighting factors and recalculate the matrix for the generic 1950s automobile.

12.4 Draw the target plot for the matrix computed in Exercise 12.3.

PART IV: FUTURE PROSPECTS

<table>
<tr><td>CHAPTER
13</td><td></td></tr>
</table>

| CHAPTER **13** | # The Future of the Automobile and Its Infrastructure |

"Not fare well, but fare forward, voyagers." — T.S. Eliot, English poet

13.1 GLOBAL TRENDS AFFECTING THE AUTOMOBILE AND ITS INFRASTRUCTURE

As is the case with the rest of the world's activities, the future of the automobile will unfold amidst a panoply of rapid changes. We begin this final chapter, in which we attempt to divine the future of the automobile and its infrastructure, with some comments on the major trends that might impact this technological evolution. We have grouped these trends into five categories for convenience of exposition: (1) system evolution and integration, (2) the structure of demand, (3) culture, (4) technology and resources, and (5) environment. In practice, of course, all of these trends will be linked in a complex system. Moreover, projections such as we are attempting here are notoriously difficult, so we view our effort not as highly predictive, but as a stimulant to further thinking on the subject.

13.1.1   Trends in System Evolution and Integration

The recent evolution of automotive technology and infrastructure illustrates two important trends that reflect fundamental changes in the structure of these systems. First, each system is becoming more complex: added information content fuels more efficient performance both economically and environmentally. An example of this evolution is the continuing increase in power per unit displacement of the modern automotive power system, combined with lower exhaust emissions per unit distance. These advances in engine technology are dependent upon sensor systems and on-board computer systems, i.e., they imply an increasingly "intelligent" and complex vehicle. A second trend is the increasing linkages between previously disparate systems, creating added complexity and information content, but again leading to increased economic and environmental efficiency. An example of this trend is the development of electronic information systems that alert drivers to bottlenecks and traffic problems in the surrounding area and suggest less congested alternative routes.

13.1.2   Trends in the Structure of Demand

As is well known, the human population of Earth is rapidly increasing. From about 5.7 billion in 1995, it is expected to rise to between 10 and 15 billion by the mid- to late-21st century. Concomitantly, the global standard of living is expected to increase. A required enabling technology in achieving an improved standard for an increased population will be vechicles and infrastructure for transportation of people and resources.

The potential for rapid growth of the global vehicle fleet, tied to the rapid standard of living improvements in Southeast Asia and elsewhere, has already been mentioned. It is predicted that the world's motor vehicle fleet will nearly double to about one billion vehicles by 2025. What has received less attention is that this growth implies as well as rapid growth in the automotive infrastructure. Much desirable residential and agricultural land may be lost, and local biodiversity and water and air quality threatened, if this infrastructure expansion is not carried out with its potential environmental impacts recognized and minimized. (One recalls, for example, the highways constructed through hundreds of kilometers of tropical forest to connect Brazil's new capital city Brazilia with the coastal population centers.) Another obvious issue is the substantial increase in energy consumption implied by vehicle and infrastructure growth. The global vehicle fleet may require adoption of new vehicle technologies (e.g., hybrid vehicles) if at least temporary scarcities and supply perturbations are to be avoided.

A confounding factor in futurist predictions is the potential for large and small wars. As this is written, the prospect for major wars appears less likely than it has been for some decades, but smaller-scale though nonetheless locally devastating conflicts continue unabated. Wars have traditionally had great influence on transportation infrastructure. The Roman roads of two millenia ago were built in part to help achieve and maintain the dominance of the Roman Empire. Germany's autobahns and the U.S. interstate highway system were justified at least in part on military grounds. Conflict today may arise from resource scarcities fueled by automotive sector demand patterns (e.g., the Gulf War was fought, at least in part, to protect access to major sources of petroleum).

The other side of the war-related infrastructure discussion relates to countries or regions that suffer heavy damage during armed conflict. After the conclusion of hostilities, the damaged infrastructure is generally rebuilt or replaced, often with a far more modern approach and improved materials. Thus, though wars are almost always undesirable and certainly unpredictable, they can, in the long term, promote infrastructure development.

13.1.3 Cultural Trends

This book has demonstrated that impressive technological evolution has occurred in the way automobiles are made and used. Among the evidence that could be cited is better fuel efficiency, environmentally preferable choices of materials, advanced manufacturing techniques, better sensor technologies and increased information processing capabilities, and environmentally superior on-board catalytic emissions reduction. These improvements have purchased time, but have not compensated for consumer behavior: a heavier, more polluting fleet mix, more driving per vehicle, more vehicles per family. Hence, the major question concerning the future impacts of the automobile must be whether, how rapidly, and in what direction the cultural status of the automobile as an icon of personal freedom and an essential expression of personal psychology will evolve.

The possibility that electronic systems will increasingly substitute for private vehicles in not just superficial ways is enticing. Telecommuting for many types of jobs has been technologically possible for decades: it has been primarily management rigidity and culture that prevented its widespread implementation, and it is increasing management flexibility that is now encouraging telecommuting to happen. What is less obvious, but potentially much more important, is the possibility that electronic connectivity will provide a sense of freedom and individuality that will culturally displace the automotive culture. It is noteworthy that among young male adolescents in developed cultures, personal computers and "surfing the net" (accessing and playing on the Internet) are beginning to replace cars as the "cool" technology. For the next generation, cruising the local drag strip may be replaced by hanging out in cyberspace—especially as continuing improvements in bandwidth and processing power make full video real-time interactive systems, perhaps characterized by a touch of virtual reality and received on a wall-sized flat screen, de rigueur. The days of the automobile as cultural icon have lasted fifty years. They may have vanished in another fifty.

A change in the cultural attitude towards automobiles does not mean that the sector will atrophy, although it may well be very different only decades from now. What it does mean is that policies to limit unnecessary use, such as automatically assessed fees for peak period use of streets, or graduated fees based on engine horsepower, will become far easier to implement.

Nonetheless, we anticipate no significant short-term reduction in the size of the automobile sector. Cultural attitudes, after all, are only one component of global demand for automobiles. The other factor in the overall demand for automobiles, and hence their supporting infrastructure, is the combination of population and income growth. Both population and increasing discretionary income in newly developed countries are trends which we anticipate will continue, at least in the short term and barring catastrophic events. More people with the discretionary income to purchase automobiles, especially in Asia and Latin American markets, will assure that demand continues strong, at least in the short term.

Accordingly, we project a world where the automobile increasingly loses some of its cultural and psychological power, but, because of other trends, demand for vehicles remains strong. This puts a continuing premium on developing environmentally preferable energy and transportation infrastructures.

13.1.4 Technology and Resources Trends

13.1.4.1 Information Management. The primary technological trend in both automobile and infrastructure design will be the use of information collection and management systems to reduce demand for other, less environmentally preferable, inputs such as energy and raw materials, and to result in increasingly sophisticated integrated transportation systems. Increasing information complexity—from on-board sensor systems and maintenance optimization systems, to traffic control systems linking on-board map units to local traffic condition sensors and routing algorithms, to national and international product and passenger routing algorithms which provide the desired level of transportation at the least economic and environmental cost—will maintain desired levels of service while minimizing concomitant environmental impacts. Here, critical advancements will occur in sensor technology, and in developing hardware and software technologies to link individual automobile activity to local and regional traffic network conditions.

13.1.4.2 The Increasing Shortage of Petroleum. As we mentioned in Chapter 9, one can foresee the eventual exhaustion of the petroleum supply, although advanced methods of locating and recovering crude oil continue to delay the inevitable. Most of the more developed countries have limited reserves such that they may be depleted within one or two decades. Depletion will be less rapid in other regions, but it seems possible that by the end of the 21st century petroleum reserves will be close to exhaustion. The result will be a forced transition to new forms of energy for motor vehicle propulsion.

An important determinant of how rapidly the transition to new energy sources can occur is the extent to which the sizable financial investment in existing energy products and systems can be reutilized rather than abandoned. Appropriate public policies might include encouraging energy suppliers to shift their investment base (e.g., into biomass plantations and processing facilities), and providing tax subsidies or other payments to facilitate this transition.

13.1.4.3 Energy Storage. Efficient energy storage systems will be increasingly important. Technological ferment in this field is quite pronounced currently, with a number of possible technologies under investigation. These range from traditional chemical energy storage systems such as lead-acid or metal hydride batteries, which may have significant environmental deficiencies such as diffusion of lead during the battery lifecycle, to unique capacitor systems based on aerogels and other new technologies, to mechanical energy storage systems based on high revolution flywheel technologies. Another potential alternative, which can either substitute for, or complement, existing energy systems, is the advent of a hydrogen infrastructure which would require safe, conveniently and safely rechargable, on-board hydrogen storage systems. Each of the alternative technologies has disadvantages and advantages. We feel quite comfortable, however, in predicting that relatively significant advances in energy storage can be anticipated in the next decade. The predominant technology, however, and the timing of its diffusion throughout the economy, will depend on the overall rate and shape of the automotive sector's evolution as well as the development of the energy technologies.

13.1.4.4 Materials Science.   Originally, automobiles were constructed out of the same materials as the carriages they replaced, primarily wood. In the post-war period, wood increasingly gave way to steel. The search for appropriate new materials, and their integration into the built product, will be primarily driven by three considerations: strength, light weight, and lifecycle management infrastructure.

The requirements of strength and light weight will favor heavy use of composite materials (as has already happened with aircraft). The composition of such materials will cover a broad chemical range, but fiber-reinforced plastics will be among the early choices. Composite materials will permit a host of new engineering approaches, including the continuous monitoring of the condition of vehicle components and the active tailoring of such features as vehicle ride and passenger compartment environment. Light metals such as magnesium may also find numerous uses. Durable ceramic engines may last twice as long as today's 150,000 km versions, although the capability to upgrade efficiency and performance over this period will have to be assured, probably through sophisticated modular design.

Materials science is also beginning to be applied to recycling, promising an eventual flow of recycled material with advantageous properties and costs. It is important to remember that, overall, the existing recycling systems and technologies are relatively primitive: research and development efforts and technological innovation have tended to focus on product and service development, not recycling. Accordingly, as the economic and policy importance of the recycling sector accelerates, we anticipate rapid development of new systems and technologies. Accordingly, we regard the lack of existing efficient recycling technologies and systems as representative of past underinvestment, not future potential.

In practice, the selection of environmentally appropriate materials that meet other design criteria will continue to be a difficult task, frequently involving a number of trade-offs. We have presented several methodologies for making such decisions in this volume, but can confidently predict that this will continue to be an area of extensive R&D activity.

13.1.5 Environmental Trends

As discussed elsewhere in this book, the production and use of automobiles, with the associated environmental impacts of the petroleum and infrastructure construction industries, have had a number of major impacts on the environment. These are manifest at virtually all temporal and geographical scales, ranging from the diffusion of heavy metals—lead from gasoline, cadmium from tires—to numerous emissions affecting regional and global atmospheric chemistry. It is not necessary to project doomsday senarios of precipitous global change to recognize that a number of environmental trends associated with current patterns of automobile use will continue to worsen, regardless of what short term path is chosen.

13.1.5.1 Global Climate Change.   Climate has changed considerably over centuries or millenia throughout Earth's history, but concerns are now arising that climate changes may be proceeding extremely rapidly under the influence of humanity's activities. An example of the evidence leading to these concerns is the temperature record for the past century and a third, shown in Fig. 13.1a. These data show that about $0.5°C$ of warming has

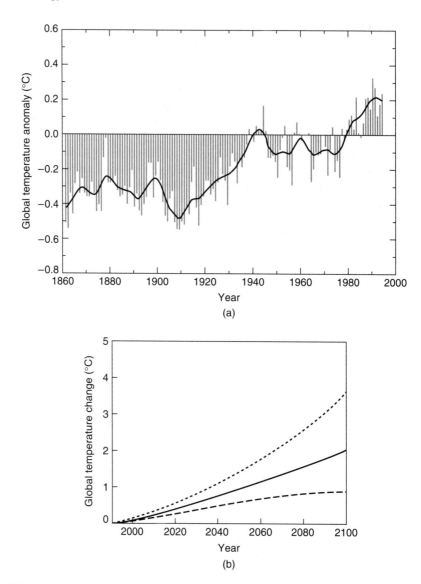

Figure 13.1   (a) Combined land-surface air and sea surface temperatures (°C) 1861 to 1994, relative to 1961 to 1990. The solid curve represents smoothing of the annual values shown by the bars. (b) Projected global mean surface temperature changes, 1990–2100. The center curve is the "best estimate"; the upper and lower curves are thought to be the highest and lowest possible changes. (IPCC Working Group I, *Climate Change 1995—The Science of Climate Change*, Cambridge, UK: Cambridge University Press, 1996.)

occurred in a hundred years, and that the most recent years have clearly been the warmest during that period. The cause is thought to be the greenhouse gases added to the atmosphere in the past two centuries (but especially the past few decades); the temperature pattern of Fig. 13.1a roughly parallels that of the use of fossil fuels.

While it is not yet absolutely certain that the observed warming is related to human activities, unquestioned evidence is likely to emerge in the first decade of the 21st century. As a result, there will be great pressure to reduce emissions of greenhouse gases, including carbon dioxide, methane, and HFC-134a. The first two are, of course, products of the combustion of fossil fuels, and the last is the coolant used in modern automotive air conditioners. These reductions will need to be accomplished in the face of a probable large expansion in the number of motor vehicles in use.

What is the anticipated global climate for the next few decades or centuries? Projections based on the most likely world development scenarios have been completed, and some of the results are shown in Fig. 13.1b. The "best estimate" value for the temperature increase by 2100 of about 2°C. Optional choices for the pace of development and for the response of climate to greenhouse forcing provide a range of 0.9–3.5°C. An average global temperature change of a few degrees does not seem very large until the change is put into perspective with past climate oscillations. For example, the Little Ice Age of 1400–1650 was about 0.5°C cooler than the present, and forced significant changes in agricultural practice and habitation. Should warming of one or two degrees occur, the planet would be at a temperature probably not seen for 120,000 years.

13.1.5.2 Sea Level Rise.   A probable consequence of global climate change is an increase in the average level of the oceans. Computer models predict that the change in global mean sea level will be about 15 cm by the year 2030 and about 50 cm by 2100, mostly due to thermal expansion of the oceans and the increased melting of mountain glaciers (Fig. 13.2). The limits of uncertainty give a range to the 2100 estimate of 13–94 cm. Should a change of this order of magnitude occur, it will cause major disruptions to the lives of a large fraction of the world's population living in coastal regions, especially less developed nations in southeast Asia, and on islands. It will also stimulate major infrastructure rebuilding. This will be a huge expense, but will provide the opportunity to re-engineer a significant fraction of the transportation infrastructure.

13.1.5.3 Enhanced Frequency of Extreme Weather.   A possible consequence of global warming is an increased frequency of extreme weather events such as hurricanes, tornadoes, and heavy precipitation. Infrastructure enhancement to deal with such events on a more frequent basis may occur, and may have implications for transportation systems. This issue was faced in 1995 by Thomas Karl and colleagues at the U.S. National Climatic Data Center. Their solution was to devise a *greenhouse climate response index* to collect and represent the multifaceted nature of the entity they were attempting to study. Because climate is inherently variable, they needed to examine long-term data records, especially those related to the predictions of global climate computer models. The four indicators chosen were:

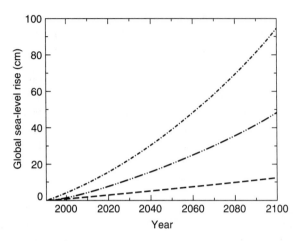

Figure 13.2   Projected global mean sea level rise, 1990–2100. The center curve is the "best estimate"; the upper and lower curves are thought to be the highest and lowest possible changes. (IPCC Working Group I, *Climate Change 1995—The Science of Climate Change*, Cambridge, UK: Cambridge University Press, 1996.)

- The percentage of the United States with much above normal minimum temperatures. (Minimum temperatures are expected to be more sensitive indicators than maximum temperatures.)
- The percentage of the United States with much above normal precipitation during the months October through April (the cold season).
- The percentage of the United States in extreme or severe drought during the months May through September (the warm season).
- The percentage of the United States with a much greater than normal proportion of precipitation derived from extreme 1–day precipitation events (i.e., those exceeding 50 mm).

Karl and colleagues defined "much above normal conditions" as those falling in the upper 10% of the century-long record. The idea is that, on average, such conditions should occur 10% of the time, year after year, if there is no significant climate change during the 20th century.

The greenhouse climate response index for 1900–1994 is shown in Fig. 13.3. Since 1978, the index has averaged 2.8% above the average of the previous years of the century, a trend that is suggestive of a climate responding to global warming. The smaller peaks in the 1930s and 1950s are largely due to drought; that of the last two decades is largely due to increasing precipitation, especially in extreme events. The data do not prove a climate response to greenhouse warming, but certainly offer supporting evidence for such a cause and effect relationship.

13.1.5.4 Ozone Depletion.   In Chapter 10, we discussed the depletion of stratospheric ozone by chlorine from refrigerants and other sources. Those species have been

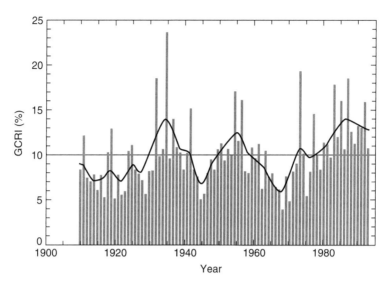

Figure 13.3   The U.S. greenhouse climate response index for the past century. The wide line is derived by averaging the value for any given year with the five years before and after. (T.R. Karl, R.W. Knight, D.R. Easterling, and R.G. Quayle, Trends in U.S. climate during the twentieth century, *Consequences, 1* (1), 3–12, 1995.)

under production and emission restrictions for two decades, and concentrations are beginning to drop fairly rapidly, as seen in Fig. 13.4. Substantial amounts of chlorine-containing species remain in the atmosphere, however, and the level of chlorine is not expected to drop below that at which the Antarctic ozone hole appeared until after the year 2050. As a result, there will be continuing pressure to avoid the use of CFCs and other halogenated species in automobile cooling systems, and to make every effort to eliminate the leakage or discharge of those amounts already present.

13.1.5.5 Habitat Destruction.   The primary present cause of endangerment or extinction of species is loss or disturbance of natural habitats. These habitats have always varied as a function of climate and other external factors, but are now disappearing at a very rapid rate due to expanded urban areas, expanded crop production, and the like. Current data suggest that Earth's primary tropical forests could be essentially gone in four or five decades. Since these forests contain the richest diversity of species on the planet, their loss would be a major blow to efforts to preserve species diversity. It is estimated that 0.2 to 0.3 percent of all species in the forests are lost per year at the present rate of deforestation; this constitutes the loss of about 4000 to 6000 species annually, perhaps 10,000 times the average natural extinction rate. The bulk of the tropical forests are in countries or regions expected to rapidly increase their numbers of motor vehicles in the next quarter century: Brazil, Indonesia, and central Africa, especially.

Loss of habitat tends to occur piecemeal. It is a difficult impact to deal with, because not only a suitable total amount of habitat needs to be preserved, but large contiguous tracts are needed for species and ecosystem survival. Once a major change in land use has

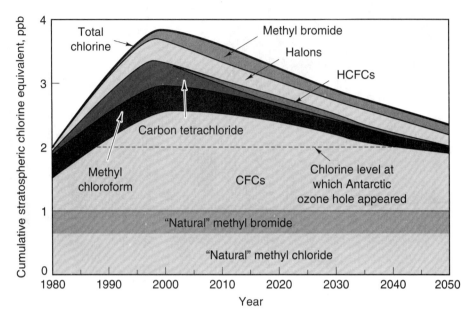

Figure 13.4 Levels of stratospheric chlorine equivalents measured (1980–1995) and anticipated (1995–2050) under the provisions of the Montreal Protocol on Ozone Layer Depletion. The ozone-depleting effects of the bromine atoms in halons and in methyl bromide have been converted to their chlorine equivalents. (Courtesy of M. McFarland, DuPont Corporation.)

occurred, the habitat is irretrievably lost, because survival of native plant and animal life is no longer possible. Even where habitat is retained, disturbances caused by nearby human activities may have significant negative impacts. Habitats near urban areas where roadway networks are enhanced will be under particular pressure in the next few decades. New roads that open access to previously difficult to reach areas, such as virgin tropical rain forests, will inevitably result in habitat degradation. Such constructions should be avoided where possible, and only undertaken where an objective assessment demonstrates that the benefits outweigh the environmental costs.

13.1.5.6 Photochemical Smog.   The trends of the incidence of smog and visibility impairment in the major cities of the world are not encouraging. A recent survey showed 20 of 24 metropolitan areas with populations over 10 million people had levels of air pollution such that human health could be affected. The most common problem was suspended particulate matter, the next the concentration of ozone. Both are related to emissions from automobiles as well as to industrial activity.

As the number of automobiles and the levels of industrial activity continue to increase in the developing countries, there will be strong pressure to minimize automotive emissions. Provided the costs can be borne, many cities, including Bangkok, Calcutta, Manila, and Tehran as well as Mexico City, London, and Osaka (to choose a few large cities at random)

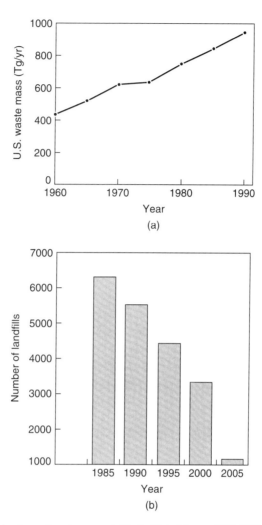

Figure 13.5   (a) The total volume of wastes generated in the United States, 1960–1990. (b) The number of landfills in the United States. For 1985 and 1990, actual figures are given; the later years are projections made by the U.S. EPA. (Drawn from data published by the U.S. Environmental Protection Agency, 1993.)

can be expected to limit automotive traffic, implement public transportation systems, impose alternative propulsion systems, or combinations of all these methods.

13.1.5.7 Waste Disposal.   The amount of solid wastes being generated on the planet continues to increase, as seen in Fig. 13.5. At the same time, the number of landfills (though perhaps not total landfill volume) is rapidly decreasing in many parts of the world. The combination of these two factors has driven governments to address problems of proper disposal and, importantly for our discussion here, of avoiding or at least minimizing the generation of wastes in the first place. The costs of landfilling wastes, already high in many regions, is likely to continue to increase. More and more, however, disposal will not be a permitted option, and industries and individuals will be required to find alternatives. Redesigning ASR to be recycled for materials and energy recovery would be an important contribution to relieving these stresses, and will most likely be driven by policies in regions such as Europe or Japan where land is becoming increasingly scarce.

13.2 THE PERSONAL TRANSPORTATION VEHICLE OF THE FUTURE

13.2.1 Goals for New Vehicle Designs

Designers of the vehicles of the future are well advised to start without preconceived notions concerning their designs, and then work down to what can be done in the short term. Useful guidance is provided by identifying what the goals of those designs should be. We propose that they include the following:

- Reduce the energy intensity of the typical automobile
- Reduce the environmental impacts of the typical automobile during its use
- Reduce the automotive industry's consumption of materials resources, especially where materials in short supply are concerned.
- Improve the crash protection of vehicle occupants
- Optimize the modularity and recyclability of the typical automobile

13.2.2 Characteristics of Future Vehicles

13.2.2.1 Classes of Vehicles. Engineers, like the public at large, are psychologically constrained by existing technologies. The first cars looked like horse-drawn carriages; the first television programs were stage performances that were photographed. If the bicycle were to be invented today, people would probably try to make it look and act like a car. Thus, a proverbial Martian might well ask, on seeing our automotive sector in action, "Why are so many people driving around alone, but hauling around space for six?" One simple and not altogether incorrect answer is that the typical buggy of a century ago could transport more than one or two passengers.

As population increases over the next few decades, and congestion becomes even more common than it is today, the average size of personal transportation vehicles will inevitably decrease. Tradition has it that vehicles that are small, and perhaps hold few passengers, are "inferior cars." It is true that small vehicles do not have the same characteristics as large vehicles, but this need not necessarily make them inferior. Consider some of the differences: small vehicles hold less luggage than large vehicles, they hold fewer passengers, they burn less fuel, and they fit into a greater variety of parking spaces. One could draw up a similar list for other types of vehicles currently on the road such as motorcycles or bicycles. No one expects that a motorcycle will carry several passengers or that a bicycle

a

b

c

Figure 13.6   Examples of new classes of personal vehicles produced as "concept cars" by different manufacturers. (a) The General Motors Ultralite, a four-passenger, gasoline-efficient sedan; (b) The Ford Motor Company Clamshell, a 2+1–wheeled vehicle seating two passengers in tandem; (c) The Quincy-Lynn Trimuter personal transportation vehicle.

will carry lots of luggage—they are simply personal transportation vehicles designed to do a different job from that for which the traditional large automobile is designed. One can therefore picture the creation of a variety of vehicles for a variety of uses.

Let us define three possible classes of future personal transportation vehicles:

The *touring car* is a lighter and perhaps slightly smaller version of today's typical automobile (The GM Ultralite of Fig. 13.6a is an example). It is designed to carry several passengers and a reasonable amount of luggage while consuming modest amounts of gasoline. The touring car is primarily suitable for long-distance transport.

The *suburban car* is designed largely for modest-distance commuting and similar uses. The normal capacity is two passengers and a small amount of cargo. Early versions of the suburban car would be capable of long-distance travel, but would not customarily be used for it. An example of a current suburban car design is the Ford Clamshell (Fig. 13.6b). Later versions will grow into a different kind of vehicle—smaller, light, probably electric—perhaps not unlike a ruggedized, upgraded golf cart.

The *city car* is designed for short-distance transportation, probably by only a single person. It is small enough to be easily maneuvered and parked in congested urban areas. An example is the Quincy-Lynn Trimuter (Fig. 13.6c).

What are the prospects for public acceptance of the suburban car and the city car? Prospects may be good if the vehicles can be perceived as alternative means of personal transportation, much as motorcycles or bicycles are now, rather than as cramped touring cars. Part of the design approach will be to produce suburban cars and city vehicles that are seen as stylish and luxurious, not as cheap touring car alternatives. In a real sense, they will not be cars at all, but a completely different product, as trucks and golf carts are today. A high degree of electronics to aid in urban navigation may help to position city and suburban vehicles as upscale status symbols.

13.2.2.2 New Materials.   Tomorrow's automobiles will rely very heavily on newly-developed materials, both to improve reliability and longevity and to reduce weight. Detailed predictions regarding the materials of the next century are not possible, of course, but we can point out materials whose use is sure to increase. They are as follows:

- Engineered plastics. Plastics use in automobiles, which has grown rapidly in the past decade, will continue as new molding and processing techniques continue to lower the cost and improve the performance of the materials. Gains are anticipated in physical strength, dual-polymer injection molding technology, stiffness, and processing time. The result could be a virtual elimination of non-plastic materials from automotive interiors, and a doubling of the total quantity of plastics used per vehicle.

- Composite materials. The most common composites, already in use in airplanes, helicopters, and automobiles to a moderate degree, are inorganic and organic fibers that are woven together, impregnated with a plastic resin, and hardened. The materials are typically lighter than metals by a factor of two to six, and can be much stronger. In addition to solid structures, honeycomb and foam composites will find a number of uses.

- Ceramics. Ceramic materials are light, highly resistant to heat, and their properties tend not to degrade with time. As a result, they are beginning to see use as engine components and for other high-temperature applications. Such uses will only increase as the ability to form new varieties of ceramics increases and as the tendency for brittle fracture is overcome by revised ceramic formulations.
- Light metals. Aluminum already sees substantial use in automobile frames, support members, and engine components. Magnesium is often used to make complex cast parts. These metals will gradually supplant more and more applications now filled by steel.

13.2.2.3 Propulsion Systems.   As discussed in Chapter 9, a number of options now exist for moving from an engine powered solely by gasoline to one using an alternative propulsion system. The first such system is now on the road, though not yet common: the electric car. Such vehicles are, however, inefficient (they must haul around hundreds of kilograms of lead to provide themselves with energy) and are flawed conceptually. An all-electric vehicle of today should be seen as a city car; the effort to make it a touring car is technologically oxymoronic. Soon to follow will be the hybrid vehicle, a combination of gasoline and electric propulsion. Over the longer term, vehicles with progressively smaller environmental impacts will appear. For eventual transition to a vehicle powered by fuel cells, for example, a programmed three-step process (Fig. 13.7) is probably a much sounder approach than a jump directly from the internal combustion engine, for both engineering and economic reasons.

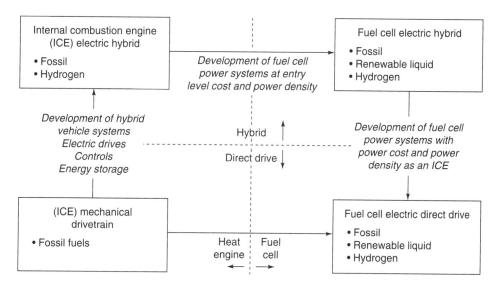

Figure 13.7   The evolution of vehicle propulsion systems from gasoline-powered to fuel cell-powered.

13.2.2.4 Electronics. The modern automobile, already properly pictured as a "computer on wheels," will increasingly become electronically controlled. Among the characteristics now being developed or readily seen to be feasible are the following:

- Electronically-controlled suspension systems that can be changed on command to provide a sporty or comfortable ride to order, or anticipate bumps and road variations and instantaneously adapt to them to smooth the ride.
- Continuous evaluation of vehicle operation, with rapid-response sensors such as radar-based electronic dipsticks for monitoring fluid levels and strain gauges for monitoring tire pressure.
- Intelligent cruise control, in which sensors determine how much distance to maintain from the vehicle ahead and automatically hold that distance rather than a preselected speed.
- An electronic road atlas, which uses global positioning system (GPS) satellites to locate the vehicle precisely and then suggests to the driver a route to the destination that is efficient and avoids known traffic problems.
- Electronic steering, accelerating, and braking upon response from a single joy-stick controlled by the driver. Electronic systems will manage energy recycling by using braking to generate reserve power.
- An advanced vehicle control system that, under general instruction from the driver, uses digital video cameras mounted front and rear to sense the road, nearby vehicles, directional signs, and other information to send suitable control instructions to steering systems, accelerator, and brakes.

Additional electronics for safety, convenience, communications, and driver information are also assured. The speed with which these controls can be incorporated into vehicles is probably not dependent very much on progress in electronics; it will rest more squarely on devising complex electronic systems that nonetheless are straightforward for a variety of human operators to use, bringing those innovations to market at an acceptable cost, and developing the supporting technological infrastructure (e.g., intelligent roadway systems) in a timely fashion.

13.2.2.5 Modularity. Today's automobiles have a modest degree of modularity in that defective parts and components can be removed, repaired, and replaced, but often not easily. In the future, it is likely that engines, transmissions, suspension systems, electronics, body components, and other parts will be designed so that they can be removed and replaced as easily as can today's portable radio battery. Such modularity will dramatically decrease the amount of time vehicles are unavailable due to maintenance and will make it more likely that defective parts will be retained and refurbished within the existing automotive infrastructure system. It will also allow for both technological advances—with concomitant increases in economic and environmental efficiency—and product life extension, thereby reducing the velocity of materials flow through the automotive sector. The tradeoff is that modularity tends to lead to somewhat overdesigned components or to somewhat undersized performance.

13.2.2.6 Driver and Passenger Safety.   The continued electronics revolution in automobiles will contribute to the safety of drivers and passengers as well as to vehicle control and performance. Small vehicle radars that look forward, back, and to the sides will provide warning of impending collisions and, ultimately, will automatically brake or steer so as to avoid or minimize the crash.

Sensors are also being developed to monitor the drivers of the vehicles, to alert tired drivers by audible signals, and to deny vehicle operation to drivers who are incapacitated by medical emergency, intoxication, or other infirmity.

New materials enter into the safety picture by virtue of their strength and flexibility. For example, fiber-reinforced plastics can absorb more than five times the energy than can steel, and they thus appear particularly suitable for use in automotive components subject to crash deformation such as side body panels and front ends.

13.2.2.7 Multi-Platform Vehicles.   Today's approach to a vehicle purchase is to determine the principal use for the vehicle and to select a vehicle appropriate for that use, with perhaps a small amount of consideration to secondary uses. Thus we find a buyer purchasing a convertible as a young adult, a station wagon as a married adult, and a sedan as an older person. If needs change (twins suddenly arrive, for example), the vehicle is sold and another is purchased. The result is frequent vehicle turnover, plus a tendency to purchase a vehicle larger than usually needed because of the occasional requirement for a vehicle capable of transporting more than the usual number of people or cargo.

An alternative approach is to design a vehicle so that different bodies can be quickly and easily put on the same chassis. Mercedes-Benz, which is furthest along in thinking this idea through, is evolving a concept in which the same basic chassis could be converted in a fraction of an hour into a sedan, convertible, station wagon, or pickup (Fig. 13.8).

13.3 THE TRANSPORTATION INFRASTRUCTURE OF THE FUTURE

13.3.1   Goals for Future Transportation Infrastructure

As with the automobile of the future, the transportation infrastructure of the future (including the roadway system and other system components) cannot be designed until its goals are carefully identified. A reasonable list might include the following:

- Reduce the energy intensity of the transportation infrastructure
- Reduce the environmental impacts of the transportation infrastructure
- Reduce system inefficiencies such as traffic jams, toll booth lines, lane closings as a result of accidents, etc.
- Reduce the rate of consumption of resources for infrastructure development and maintenance
- Improve the accident avoidance characteristics of the infrastructure-automobile system

210

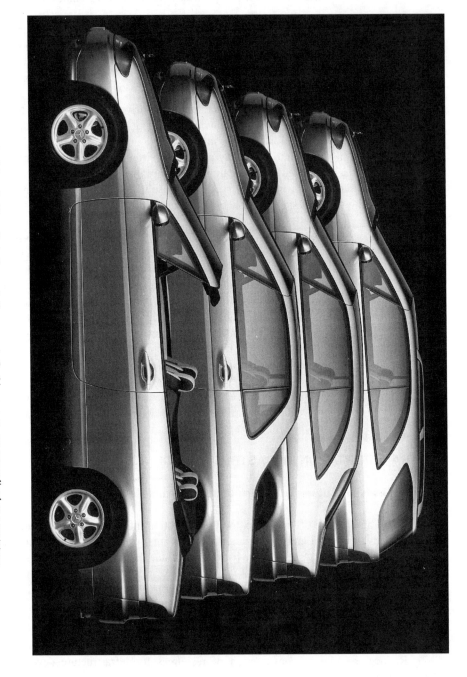

Figure 13.8 The Mercedes/Benz Vario Research Car concept. In this concept, not yet realized as a prototype, the same chassis can be quickly converted to any of several types of vehicle. (Courtesy of Mercedes-Benz AG, Stuttgart.)

- Improve the utility of the infrastructure-automobile system in severe weather conditions

- Migrate transportation users to the most efficient transportation mode for their needs, and make sure such services are available where and when needed.

- Plan transportation infrastructures so that they are linked to other infrastructure systems, e.g., housing and manufacturing, in such a way as to optimize the environmental and economic efficiency of the societal technology system as a whole.

13.3.2   Characteristics of Future Transportation Infrastructure

13.3.2.1   The Electronic Highway.   The first steps are now being taken toward the electronic highway, an infrastructure designed to optimize the flow of traffic and minimize congestion and accidents. The electronic toll booths that today in a few locations permit drivers to pay fares merely by passing through a specified section of roadway at reduced speed will eventually evolve to a system in which the communication between the infrastructure and the vehicles traveling on it will be extensive. Drivers will be actively warned of unsatisfactory weather conditions, traffic tie-ups, and accidents, and alternative routes will be furnished to on-board computers in the vehicles. Tolls will be collected automatically at the routine traveling speed. If desired, information can be transmitted about tourist attractions, housing, and refueling. In addition, the congestion potential of central urban regions and other critical transportation nodes will be extensively controlled by road-pricing approaches that will charge very high rates for vehicles wishing to travel at peak traffic times.

A further concept that may eventually be realized will be active control of vehicles on electronic highways by the highway infrastructure rather than by the driver. This can be thought of as akin to today's practice of giving over control of a vehicle to the operator of a train or ferry onto which you have driven your automobile. On the electronic highway, distances between vehicles, lane selection, and exiting would be handled by the infrastructure itself, permitting much higher rates of vehicle transit than would be the case if each individual vehicle were controlling itself for its own purposes (Fig. 13.9).

13.3.2.2   Alternative Forms of Personal Transportation.   It seems inevitable that the concept of personal transportation that relies primarily on a single type of vehicle will gradually evolve into a system with the ubiquitous mix of alternatives now seen in a few prosperous, well-designed cities—personal vehicles, rail transport, air transport, bicycles, pedestrian walkways. The goal will be to permit personal flexibility while enhancing the efficiency of the transportation infrastructure system. The personal vehicle will become not the sole means of transport, but a part of a larger structured system that optimizes each part rather than placing the ultimate responsibility for travel within a single type of vehicle.

13.3.2.3   Community Vehicles.   Although as a society we have evolved into a pattern where those who want to drive cars need to own them, different scenarios may be imagined. One contemplated by General Motors is a customer purchasing the right to the

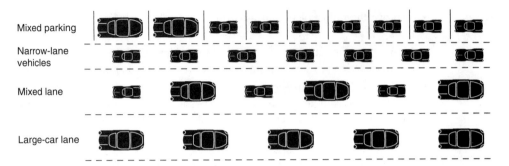

Figure 13.9 A concept of mixed-size vehicle transport on an electronic highway. (Adapted with permission from R.Q. Riley, *Alternative Cars in the 21st Century*, Warrendale, PA: Society of Automotive Engineers, 1994.)

services of a vehicle rather than the vehicle itself. The customer might then have a vehicle on long-term assignment to her or him while at home, but be given temporary rights to a vehicle while traveling on business in San Francisco or on vacation in Greece or while moving a young adult from college to an apartment. The vehicles need not all be the same; some uses suggest convertibles, some trucks, some sedans. The concept is, in reality, a sort of juxtaposition of car leasing, car rental, and truck rental under a single contract where changes in needs could be handled on a per-use basis much as banks charge for each use of automatic teller machines. The benefit to the user is enhanced flexibility. For the corporation, it is an enhanced relationship with the customer. From an environmental standpoint, if the ownership of the vehicles remains with the corporation, the chances of components, materials, and vehicles themselves being maintained, recycled, and reused increases dramatically. Such systems approach the European concept of a "functionality economy," in which consumers purchase function, not underlying product, from the manufacturer, who thus becomes responsible for the environmental impacts attributable to production, use, and recycling of the product.

Vehicle sharing can be viewed in a broader context as well. Large metropolitan areas follow a familiar pattern; every morning commuters drive vehicles into the city and park them in surface lots or parking structures. Meanwhile, employees whose jobs require mobility—taxi drivers, package deliverers, sales and service personnel—are taking vehicles from other parking lots and beginning their business days on the streets. The system seems tailor-made for organizing into one with fewer vehicles, but vehicles that are well-maintained and better integrated into the municipal transportation infrastructure.

An example of evolving patterns in automobile ownership and use is provided by the VivallaBil car co-op in Orebro, Sweden. This organization is made up of twenty-five households that among them own a half-dozen automobiles. Reservations must be made for use of an automobile, a daily charge is assessed, and the vehicle must be returned with a full tank of gasoline. Several additional automobiles are rented by the co-op in the summer, when demand is higher. VivallaBil co-op members are more likely to want vehicles in the evening and during normal commuting times. Thus, during midday some of the vehicles are rented to the municipality, which uses them for community business and thus avoids buying its own fleet. The result of the co-op is that members plan their activities more and drive less. Those involved have clearly decreased their transportation expenses at the cost of some inconvenience to themselves.

13.3.2.4 City and Suburban Planning.

Transportation systems work well only within regions that are planned with transportation as one of the central factors. The single most difficult attribute of North American urban/suburban development from the standpoint of efficient transportation has been the tendency for housing, shopping, and businesses to be located without regard for concentration or for utilizing the existing infrastructure of trains, buses, and walkways. Once any geographically dispersed use is established, collecting people into any efficient transportation network becomes expensive and difficult.

Visionary urban planners thus see an evolution to metropolitan areas that are developed around a network of alternative and efficient transportation. Without sacrificing individual preferences and choices, providing the opportunity for pleasant housing and easy shopping and commuting in a variety of ways can be enabled by designing core residential and commercial areas around transportation corridors. Personal vehicles will likely be important for the first leg of any trip, but public transportation of several types would be relied upon thereafter. Safety, convenience, and cost would all need to be addressed in such a picture, but it is eminently clear that urban planning in the 21st century must encompass three factors relevant to this book and to the transportation infrastructure: moving people, moving resources, and substituting information for transportation wherever feasible. Unless those factors are addressed in detail and with vigor and commitment, metropolitan areas will become increasingly difficult places to live.

13.4 SUMMARIZING THE FUTURE VISION

The automobile of the future will be more complex at all levels, with each function controlled and linked with others through a complex system of sensors and feedback loops. All unnecessary loss of energy in operation will be eliminated: for example, flywheels will capture energy now lost through braking, and excess engine heat will be recaptured chemically for reuse. Sensor systems will minimize maintenance impacts, requiring, for example, replacement of engine oil only when indicated by the actual condition of the oil and the engine. Unlike today's autonomous artifact, the automobile of the future will be up-linked into its local and regional transporation infrastructure for a number of purposes, including automatic assessment of road use fees based on time of day, route, and engine efficiency. The driver will benefit from downloaded information providing realtime maps and directions, and information on traffic and road conditions. This information will allow him or her to operate the automobile more efficiently.

Additionally, modular automobile design will become more popular. Today's automobiles, designed to carry at least four people and substantial luggage but frequently only carrying the driver, will be seen as extremely wasteful, and replaced by personal transportation units that can add passenger and luggage space as required. Accordingly, especially as the automobile is displaced by electronic communications devices as the symbol of personal freedom, the line between private and public transportation will begin to blur. Concomitantly, consumers will learn to think of function rather than artifact, demanding different forms of transport rather than making a single artifact meet all needs.

The infrastructure supporting the automobile will also change radically. Like the automobile itself, it will become "smart," substituting information management and communication for energy and resource consumption. Analogously to electricity, demand for new roadways will be managed not by construction, but by "demand side management," and conservation of roadway availability will be stimulated by pricing mechanisms such as high fees for travel during peak hours. Sensor and information management technology will evolve to support such a system. Increasingly, the information infrastructure, not the physical infrastructure, will be critical to achieving improved environmental and economic performance of the automobile during its use lifecycle phase.

The energy sector will clearly undergo considerable change in response to altered automobile design. Although we can predict decarbonization of energy production for automotive consumption with some certainty, the form of energy that will eventually predominate—methane or natural gas derived from petroleum, coal or biomass, or electricity derived from sustainable fuels, including perhaps nuclear or fusion power—is less easy to foresee.

The construction sector will also evolve considerably. Reflecting trends affecting other products, the industry will increasingly concentrate on "product life extension"—that is, developing materials and technologies by which the active life of infrastructure can be lengthened. Additionally, it will be increasingly the case that residue streams from both new construction and demolition of old construction will be recycled, either voluntarily or as a result of regulation. As of January 1996, for example, The Netherlands implemented a ban on landfilling construction waste. Among other things, these trends will stimulate significant increases in research and development activity in this sector, which has historically been very slow to incorporate new materials and technologies.

Society will increasingly substitute electronic products and information management for automobiles, both directly (e.g., teleworking), and culturally. This transformation will facilitate the evolution of an "automotive functionality economy", where consumers purchase the use of an automobile, rather than the automobile itself. A variety of leasing, sharing, and rental arrangements will support this method of obtaining transportation. The dominant automobile manufacturing firms will restructure to become "transportation specialists" rather than just manufacturers, and will obtain competitive advantage through attractive service packages, rather than from product design and manufacture. Indeed, the seeds of this evolution are being planted, perhaps unintentionally, in the rapidly expanding leasing programs offered by most automobile manufacturers, and the vertical integration of some manufacturers into the automobile rental service sector.

13.5 MOVING FROM HERE TO THERE

Given that society can assess where it now is, and has a picture of the transportation future it wishes to achieve, how should it manage the transformation from the system of today to that of tomorrow? The key ingredient in enabling the transformation is to think first of the need that society desires to fill, not of the objects that have been created to fill similar needs in the past and present. In other words, planners should ask where people wish to travel and what amenities they wish as they travel, and then design systems on local, regional, national, and international scales to meet those needs while having minimal impacts on human societies and natural and planetary ecosystems. Cultural, non-transport needs currently met by automobiles—status and wealth statements, self-image reinforcement—must be considered in this process. Innovation will be the key; tomorrow's transportation system need not resemble today's. Timing and technological development will define some of the options, but it seems likely that a mix of small personal vehicles, larger family vehicles, rail, and air transport will be part of the optimum solution, and those pieces will need to be linked together to optimize the sequencing among them.

Once the desired systems are identified, environmentally-sensitive engineering will be needed to design them and make them efficient and robust. A high degree of sensor provisioning and electronic control will be needed, and backups and failsafe designs will be required. The phasing in of the different parts of the system on a carefully planned schedule will be crucial to a successful transition. The implementation of such a system will require a high degree of public support, across political boundaries ranging from local to global. This societal part of the transition may involve modification of zoning laws, the redesign of international transportation standards, and a modification of tax incentives and disincentives. Cooperative international planning will be crucial (perhaps building on the model of ground and air transport systems in the Scandinavian countries).

The transition from the present system, one that is complex, intertwined, mostly disconnected, and involving both private and public actors, into a system that provides improved transport in new or modified ways will be difficult, and aspects of it will be expensive. Accomplishing that transition will be one of the supreme tests of the industrial ecology paradigm on which we have based this book. How we as a society deal with our need and desire for transportation may well control significant aspects of the future of our planet and species; at the least, it will certainly influence our happiness and that of our descendents.

SUGGESTED READING

Banerjee, T., and M. Southworth, Eds., *City Sense and City Design: Writings and Projects of Kevin Lynch*, Cambridge, MA: MIT Press, 853 pp., 1990.

Economist (London), Taming the Beast: A Survey on Living With the Car, 18 pp., 22 June, 1996.

Eisenberger, P.M., Ed., *Basic Research Needs for Vehicles of the Future*, Princeton, NJ: Princeton Materials Institute, 52 pp., 1995.

Karl, T.R., R.W. Knight, D.R. Easterling, and R.G. Quayle, Trends in U.S. climate during the twentieth century, *Consequences, 1* (1), 3–12, 1995.

Kassakian, J.G., H.C. Wolf, J.M. Miller, and C.J. Morton, Automotive electrical systems circa 2005, *Spectrum, 33* (8), 22–27, 1996.

Lovins, A.B., and L.H. Lovins, Reinventing the wheels, *The Atlantic Monthly*, pp. 75–93, January, 1995.

Mage, D., G. Ozolins, P. Peterson, A. Webster, R. Orthofer, V. Vandeweerd, and M. Gwynne, Urban air pollution in megacities of the world, *Atmospheric Environment, 30*, 681–686, 1996.

Nardis, S., and J.J. MacKenzie, *Car Trouble*, Boston: Beacon Press, 1993.

Riley, R.Q., *Alternative Cars in the 21st Century*, Warrendale, PA: Soc. of Automotive Engineers, 396 pp., 1994.

EXERCISES

13.1 Pick any one of the major environmental impacts of the automobile, and describe what the automobile and its supporting infrastructures would look like if just this effect were addressed. Then compare your results with those suggested in this book. How do they differ?

13.2 (a) Make a list of all the trips you or a representative family member take by automobile in a week. Considering only distance and purpose of trip (e.g., single person commute of ten km, overnight vacation with luggage for four), and *not* infrastructure, select the most environmentally appropriate transport system: walking, bicycling, three-wheel bike, electric golf cart, city car, suburban car, full-service sedan, four-wheel-drive sports utility vehicle. (b) What physical factors prevent you from using the most efficient option (e.g., no bike paths)? (c) What cultural factors prevent you from using the most efficient option?

13.3 Computer models predict that an average planetary warming of 2°C would raise global mean sea level by 45 cm. What would be the consequences of such a change for Bangladesh, Marshall Islands, Argentina, and Japan?

13.4 The total land area of Earth is 5.10×10^8 km^2. If tropical rain forest area in 1990 was 7.7×10^6 km^2, what fraction of the surface was tropical rain forest? What will be the fraction in 2020 if the area is decreased by 12% per decade, as predicted in some studies?

CHAPTER

14 | Epilogue

"As to the future, your task is not to foresee, but to enable it."
— Antoine de Saint Exupery, French author

In ending this book, it is appropriate to step back from the focused theme of the automobile and its infrastructure and look into the larger frame of the interactions linking technology and society. The first point to be made, an important one for engineers and technologists, is that individual decisions by designers and design teams are crucial to evolving the relationship between technology and environment. In a complex relationship such as this, some decisions may eventually turn out not to be ideal, but most of them will, and little by little the sum of those decisions will help clear the air and water, minimize the use of resources, and enable the end of the "disposable society".

No matter how "green" the engineer's products may be, however, they may well not add up to a green society. The counterexample has been before us in this book: automobiles that are more and more environmentally responsible each year, but a society that promotes the manufacture of ever-increasing numbers of vehicles and provides an infrastructure to enable them to be used more extensively than ever before. The result is a personal transportation system characterized by high technology, heavy use, and frequently inadequate performance.

Societal structures that inherently involve technology—technological-societal systems (TSS)—are common in modern society. Automobiles and related infrastructure is certainly one example, but so are buildings, so is the electric power grid, so is telecommunications. We reach an optimum TSS not by optimizing the environmentally-related aspects of each system component, but by optimizing the entire system.

We can now draw a clear distinction between the Design for Environment (DFE) actions that we have advocated throughout this book, and the industrial ecology (IE) approach that has the potential to lead to sustainable development for Earth's people. IE is the comprehensive vision of where a technological society wishes to go, and how it wishes to get there. It includes engineers in its roster of implementers, but also environmental scientists, political scientists, lawyers, economists, educators. It designs the broad picture for the participants. The engineer's role, an important one, is to make the technological aspects of that broad picture work in practice, by applying DFE concepts and tools to every engineering activity. The engineer can enable or frustrate sustainable development, but she or he cannot achieve it. That achievement is society's challenge, and the involvement and cooperation of all parts of society will be needed to make it happen.

Environmentally-Responsible Product Matrix: Scoring Guidelines and Protocols

The Product Improvement Matrix is described in Chapter 12. In this appendix, a sample of possible scoring considerations appropriate to each of the matrix elements is presented. It is anticipated that different products will require different check lists and evaluations, so this appendix is presented as an example rather than as a universal formula.

Product Matrix Element: 1,1
Life Stage: Premanufacture
Environmental Concern: Materials Choice

If any of the following conditions apply, the matrix element rating is 0:

- For the case where supplier components/sub-systems are used: no/little information is known about the chemical content in supplied products and components.
- For the case where materials are acquired from suppliers: a scarce material is used where a reasonable alternative is available. (Scarce materials are defined as antimony, beryllium, boron, cobalt, chromium, gold, mercury, the platinum metals (Pt, Ir, Os, Pa, Rh, Ru), silver, thorium, and uranium.)

If all of the following conditions apply, the matrix element rating is 4:

- No virgin material is used in incoming components or materials.

If neither of the above ratings is assigned, complete the checklist below. Assign a rating of 1, 2, or 3 depending on the degree to which the product meets DFE preferences for this matrix element.

- Is the product or process designed to minimize the use of materials in restricted supply (See above list)?
- Is the product or process designed to utilize recycled materials or components wherever possible?

Product Matrix Element: 1,2
Life Stage: Premanufacture
Environmental Concern: Energy Use

If any of the following conditions apply, the matrix element rating is 0:

- One or more of the principal materials used in the product requires energy-intensive extraction and suitable alternative materials are available that do not. (Materials requiring energy-intensive extraction are defined as virgin aluminum, virgin steel, and virgin petroleum.)

If all of the following conditions apply, the matrix element rating is 4:

- Negligible energy is needed to extract or ship the materials or components for this product.

If neither of the above ratings is assigned, complete the checklist below. Assign a rating of 1, 2, or 3 depending on the degree to which the product meets DFE preferences for this matrix element.

- Is the product designed to minimize the use of virgin materials whose extraction is energy-intensive?
- Does the product design avoid or minimize the use of high-density materials whose transport to and from the facility will require significant energy use? (Such materials are defined as those with a specific gravity above 7.0).
- Is transport distance of incoming materials and components minimized?

Product Matrix Element: 1,3
Life Stage: Premanufacture
Environmental Concern: Solid Residues

If any of the following conditions apply, the matrix element rating is 0:

- For the case where materials are acquired from suppliers: metals from virgin ores are used, creating substantial waste rock residues that could be avoided by the use of recycled material, and suitable material is available from recycling streams.
- For the case where supplier components/sub-systems are used: all incoming packaging is from virgin sources and consists of 3 or more types of materials.

If all of the following conditions apply, the matrix element rating is 4:

- For the case where materials are acquired from suppliers: no solid residues result from resource extraction or during production of materials by recycling (example: petroleum).
- For the case where supplier components/sub-systems are used: either no packaging material is used or the supplier takes back all packaging material.
- For the case where supplier components/sub-systems are used: incoming packaging is totally reused/recycled.

If neither of the above ratings is assigned, complete the checklist below. Assign a rating of 1, 2, or 3 depending on the degree to which the product meets DFE preferences for this matrix element.

- Is the product designed to minimize the use of materials whose extraction or purification involves the production of large amounts of solid residues (i.e., coal and all virgin metals)?
- Is the product designed to minimize the use of materials whose extraction or purification involves the production of toxic solid residues? (This category includes all radioactive materials.)
- Has incoming packaging volume and weight, at and among all levels (primary, secondary and tertiary), been minimized?
- Is materials diversity minimized in incoming packaging?

Product Matrix Element: 1,4
Life Stage: Premanufacture
Environmental Concern: Liquid Residues

If any of the following conditions apply, the matrix element rating is 0:

- For the case where supplier components/sub-systems are used: metals from virgin ores that cause substantial acid mine drainage are used, and suitable material is available from recycling streams. (Materials causing acid mine drainage are defined as copper, iron, nickel, lead, and zinc.)
- For the case where materials are acquired from suppliers: the packaging contains toxic or hazardous substances that might leak from it if improper disposal occurs.

If all of the following conditions apply, the matrix element rating is 4:

- For the case where materials are acquired from suppliers: no liquid residues result from resource extraction or during production of materials by recycling.
- For the case where supplier components/sub-systems are used: no liquid residue is generated during transportation, unpacking, or use of this product.

If neither of the above ratings is assigned, complete the checklist below. Assign a rating of 1, 2, or 3 depending on the degree to which the product meets DFE preferences for this matrix element.

- Is the product designed to minimize the use of materials whose extraction or purification involves the generation of large amounts of liquid residues? (This category includes paper and allied products, coal, and materials from biomass.)
- Is the product designed to minimize the use of materials whose extraction or purification involves the generation of toxic liquid residues? (These materials are defined as aluminum, copper, iron, lead, nickel, and zinc.)
- Are refillable/reusable containers used for incoming liquid materials where appropriate?
- Does the use of incoming components require cleaning that involves a large amount of water or that generates liquid residues needing special disposal methods?

Product Matrix Element: 1,5
Life Stage: Premanufacture
Environmental Concern: Gaseous Residues

If any of the following conditions apply, the matrix element rating is 0:

- The materials used cause substantial emissions of toxic, smog-producing, or greenhouse gases into the environment, and suitable alternatives that do not do so are available. (These materials are defined as aluminum, copper, iron, lead, nickel, zinc, paper and allied products, and concrete.)

If all of the following conditions apply, the matrix element rating is 4:

- No gaseous residues are produced during resource extraction or production of materials by recycling.

If neither of the above ratings is assigned, complete the checklist below. Assign a rating of 1, 2, or 3 depending on the degree to which the product meets DFE preferences for this matrix element.

- Is the product designed to minimize the use of materials whose extraction or purification involves the generation of large amounts of gaseous (toxic or otherwise) residues? (Such materials are defined as aluminum, copper, iron, lead, nickel, and zinc.)

Product Matrix Element: 2,1
Life Stage: Product Manufacture
Environmental Concern: Materials Choice

If any of the following conditions apply, the matrix element rating is 0:

- Product manufacture requires relatively large amounts of materials that are restricted {see (1,1)}, toxic, and/or radioactive.

If all of the following conditions apply, the matrix element rating is 4:

- Materials used in manufacture are completely closed loop (captured and reused/recycled) with minimum inputs required.

If neither of the above ratings is assigned, complete the checklist below. Assign a rating of 1, 2, or 3 depending on the degree to which the product meets DFE preferences for this matrix element.

- Does the manufacturing process avoid the use of materials that are in restricted supply?
- Is the use of toxic material avoided or minimized?
- Is the use of radioactive material avoided?
- Is the use of virgin material minimized?
- Has the chemical treatment of materials and components been minimized?

Product Matrix Element: 2,2
Life Stage: Product Manufacture
Environmental Concern: Energy Use

If any of the following conditions apply, the matrix element rating is 0:

- Energy use for product manufacture/testing is high and less energy intensive alternatives are available.

If all of the following conditions apply, the matrix element rating is 4:

- Product manufacture and testing requires no or minimal energy use.

If neither of the above ratings is assigned, complete the checklist below. Assign a rating of 1, 2, or 3 depending on the degree to which the product meets DFE preferences for this matrix element.

- Is the product designed to minimize the use of energy intensive processing steps?

- Is the product designed to minimize energy intensive evaluation/testing steps?
- Does the manufacturing process use co-generation, heat exchanges, and/or other techniques to utilize otherwise waste energy?
- Is the manufacturing facility powered down when not in use?

Product Matrix Element: 2,3
Life Stage: Product Manufacture
Environmental Concern: Solid Residues

If any of the following conditions apply, the matrix element rating is 0:

- Solid manufacturing residues are large and no reuse/recycling programs are in use.

If all of the following conditions apply, the matrix element rating is 4:

- Solid manufacturing residues are minor and each constituent is >90% reused/recycled.

If neither of the above ratings is assigned, complete the checklist below. Assign a rating of 1, 2, or 3 depending on the degree to which the product meets DFE preferences for this matrix element.

- Have solid manufacturing residues been minimized and reused to the greatest extent possible?
- Has the resale of all solid residues as inputs to other products/processes, been investigated and implemented?
- Are solid manufacturing residues that do not have resale value minimized and recycled?

Product Matrix Element: 2,4
Life Stage: Product Manufacture
Environmental Concern: Liquid Residues

If any of the following conditions apply, the matrix element rating is 0:

- Liquid manufacturing residues are large and no reuse/recycling programs are in use.

If all of the following conditions apply, the matrix element rating is 4:

- Liquid manufacturing residues are minor and each constituent is >90% reused/recycled.

If neither of the above ratings is assigned, complete the checklist below. Assign a rating of 1, 2, or 3 depending on the degree to which the product meets DFE preferences for this matrix element.

- If solvents or oils are used in the manufacture of this product is their use minimized and have alternatives been investigated and implemented?
- Have opportunities for sale of all liquid residues as input to other processes/products been investigated and implemented?
- Have the manufacturing processes been designed to require the maximum recycled liquid process chemicals rather than virgin materials?

Product Matrix Element: 2,5
Life Stage: Product Manufacture
Environmental Concern: Gaseous Residues

If any of the following conditions apply, the matrix element rating is 0:

- Gaseous manufacturing residues are large and no reuse/recycling programs are in use.
- CFCs are used in product manufacture.

If all of the following conditions apply, the matrix element rating is 4:

- Gaseous manufacturing residues are relatively minor and reuse/recycling programs are in use.

If neither of the above ratings is assigned, complete the checklist below. Assign a rating of 1, 2, or 3 depending on the degree to which the product meets DFE preferences for this matrix element.

- If HCFCs are used in the manufacture of this product have alternatives been thoroughly investigated and implemented?
- Are greenhouse gases used or generated in any manufacturing process connected with this product?
- Has the resale of all gaseous residues as inputs to other processes/products been investigated and implemented?

Product Matrix Element: 3,1
Life Stage: Product Delivery
Environmental Concern: Materials Choice

If any of the following conditions apply, the matrix element rating is 0:

- All outgoing packaging is from virgin sources and consists of 3 or more types of materials.

If all of the following conditions apply, the matrix element rating is 4:

- No outgoing packaging or minimal and recycled packaging material is used.

If neither of the above ratings is assigned, complete the checklist below. Assign a rating of 1, 2, or 3 depending on the degree to which the product meets DFE preferences for this matrix element.

- Does the product packaging minimize the number of different materials used and is it optimized for weight/volume efficiency?
- Have efforts been made to use recycled materials for product packaging and to make sure the resulting package is recyclable and marked as such?
- Is there a functioning recycling infrastructure for the product packaging material?
- Have the packaging engineer and the installation personnel been consulted during the product design?

Product Matrix Element: 3,2
Life Stage: Product Delivery
Environmental Concern: Energy Use

If any of the following conditions apply, the matrix element rating is 0:

- Packaging material extraction, packaging procedure, and transportation/installation method(s) are all energy intensive and less energy-intensive options are available.

If all of the following conditions apply, the matrix element rating is 4:

- Packaging material extraction, packaging procedure, and transportation/installation methods(s) all require little or no energy.

If neither of the above ratings is assigned, complete the checklist below. Assign a rating of 1, 2, or 3 depending on the degree to which the product meets DFE preferences for this matrix element.

- Do packaging procedures avoid energy-intensive activities?
- Are component supply systems and product distribution/installation plans designed to minimize energy use?
- If installation is involved, is it designed to avoid energy intensive procedures?
- Is long distance, energy intensive product transportation avoided or minimized?

Product Matrix Element: 3,3
Life Stage: Product Delivery
Environmental Concern: Solid Residues

If any of the following conditions apply, the matrix element rating is 0:

* Outgoing packaging material is excessive, with little consideration given to recycling or reuse.

If all of the following conditions apply, the matrix element rating is 4:

* None or minimal outgoing packaging material is used and/or the packaging is totally reused or recycled.

If neither of the above ratings is assigned, complete the checklist below. Assign a rating of 1, 2, or 3 depending on the degree to which the product meets DFE preferences for this matrix element.

* Is the product packaging designed to make it easy to separate the constituent materials?
* Do the packaging materials need special disposal after products are unpacked?
* Has product packaging volume and weight, at and among all three levels, (primary, secondary, and tertiary) been minimized?
* Are arrangements made to take back product packaging for reuse and/or recycling?
* Is materials diversity minimized in outgoing product packaging?

Product Matrix Element: 3,4
Life Stage: Product Delivery
Environmental Concern: Liquid Residues

If any of the following conditions apply, the matrix element rating is 0:

* The product packaging contains toxic or hazardous substances such as the acid from batteries that might leak from it if improper disposal occurs.

If all of the following conditions apply, the matrix element rating is 4:

* Little or no liquid residue is generated during packaging, transportation, or installation of this product.

If neither of the above ratings is assigned, complete the checklist below. Assign a rating of 1, 2, or 3 depending on the degree to which the product meets DFE preferences for this matrix element.

- Are refillable or reusable containers used for liquid products where appropriate?
- Do the product packaging operations need cleaning/maintenance procedures that require a large amount of water or generate other liquid residues (oils, detergents, ...) that need special methods of disposal?
- Do the product unpacking and/or installation operations require cleaning that involves a large amount of water or that generates liquid residues needing special disposal methods?

Product Matrix Element: 3,5
Life Stage: Product Delivery
Environmental Concern: Gaseous Residues

If any of the following conditions apply, the matrix element rating is 0:

- Abundant gaseous residues are generated during packaging, transportation, or installation, and alternative methods that would significantly reduce gaseous emissions are available.

If all of the following conditions apply, the matrix element rating is 4:

- Little or no gaseous residues are generated during packaging, transportation, or installation of this product.

If neither of the above ratings is assigned, complete the checklist below. Assign a rating of 1, 2, or 3 depending on the degree to which the product meets DFE preferences for this matrix element.

- If the product contains pressurized gases, are transport/installation procedures designed to avoid their release?
- Are product distribution plans designed to minimize gaseous emissions from transport vehicles?
- If the packaging is recycled for its energy content (i.e., incinerated), have the materials been selected to ensure that no toxic gases are released?

Product Matrix Element: 4,1
Life Stage: Product Use
Environmental Concern: Materials Choice

If any of the following conditions apply, the matrix element rating is 0:

- Consumables contain significant quantities of materials in restricted supply or toxic/hazardous substances.

If all of the following conditions apply, the matrix element rating is 4:

- Product use and product maintenance require no consumables.

If neither of the above ratings is assigned, complete the checklist below. Assign a rating of 1, 2, or 3 depending on the degree to which the product meets DFE preferences for this matrix element.

- Has consumable material use been minimized?
- If the product is designed to be disposed of after using, have alternative approaches for accomplishing the same purpose been examined?
- Have the materials been chosen such that no environmentally inappropriate maintenance is required, and no unintentional release of toxic materials to the environment occurs during use?
- Are consumable materials generated from recycled streams rather than virgin material?

Product Matrix Element: 4,2
Life Stage: Product Use
Environmental Concern: Energy Use

If any of the following conditions apply, the matrix element rating is 0:

- Product use and/or maintenance is relatively energy-intensive and less energy-intensive methods are available to accomplish the same purpose.

If all of the following conditions apply, the matrix element rating is 4:

- Product use and maintenance require little or no energy.

If neither of the above ratings is assigned, complete the checklist below. Assign a rating of 1, 2, or 3 depending on the degree to which the product meets DFE preferences for this matrix element.

- Has the product been designed to minimize energy use while in service?
- Has energy use during maintenance/repair been minimized?
- Have energy-conserving design features (such as auto-shut off or enhanced insulation) been incorporated?
- Can the product monitor and display its energy use and/or its operating energy efficiency while in service?

Product Matrix Element: 4,3
Life Stage: Product Use
Environmental Concern: Solid Residues

If any of the following conditions apply, the matrix element rating is 0:

- Product generates significant quantities of hazardous/toxic solid residues during use or from repair/maintenance operations.

If all of the following conditions apply, the matrix element rating is 4:

- Product generates no (or relatively minor amounts of) solid residues during use or from repair/maintenance operations.

If neither of the above ratings is assigned, complete the checklist below. Assign a rating of 1, 2, or 3 depending on the degree to which the product meets DFE preferences for this matrix element.

- Has the periodic disposal of solid materials (such as cartridges, containers, or batteries) associated with the use and/or maintenance of this product been avoided or minimized?
- Have alternatives to the use of solid consumables been thoroughly investigated and implemented where appropriate?
- If intentional dissipative emissions to land occur as a result of using this product, have less environmentally harmful alternatives been investigated?

Product Matrix Element: 4,4
Life Stage: Product Use
Environmental Concern: Liquid Residues

If any of the following conditions apply, the matrix element rating is 0:

- Product generates significant quantities of hazardous/toxic liquid residues during use or from repair/maintenance operations.

If all of the following conditions apply, the matrix element rating is 4:

- Product generates no (or relatively minor amounts of) liquid residues during use or from repair/maintenance operations.

If neither of the above ratings is assigned, complete the checklist below. Assign a rating of 1, 2, or 3 depending on the degree to which the product meets DFE preferences for this matrix element.

- Has the periodic disposal of liquid materials (such as lubricants and hydraulic fluids) associated with the use and/or maintenance of this product been avoided or minimized?
- Have alternatives to the use of liquid consumables been thoroughly investigated and implemented where appropriate?
- If intentional dissipative emissions to water occur as a result of using this product, have less environmentally harmful alternatives been investigated?
- If product contains liquid material that has the potential to be unintentionally dissipated during use or repair, have appropriate preventive measures been incorporated?

Product Matrix Element: 4,5
Life Stage: Product Use
Environmental Concern: Gaseous Residues

If any of the following conditions apply, the matrix element rating is 0:

- Product generates significant quantities of hazardous/toxic gaseous residues during use or from repair/maintenance operations

If all of the following conditions apply, the matrix element rating is 4:

- Product generates no (or relatively minor amounts of) gaseous residues during use or from repair/maintenance operations.

If neither of the above ratings is assigned, complete the checklist below. Assign a rating of 1, 2, or 3 depending on the degree to which the product meets DFE preferences for this matrix element.

- Has the periodic emission of gaseous materials (such as CO_2, SO_2, VOCs, and CFCs) associated with the use and/or maintenance of this product been avoided or minimized?
- Have alternatives to the use of gaseous consumables been thoroughly investigated and implemented where appropriate?
- If intentional dissipative emissions to air occur as a result of using this product, have less environmentally harmful alternatives been investigated?
- If product contains any gaseous materials that have the potential to be unintentionally dissipated during use or repair, have the appropriate preventive measures been incorporated?

Product Matrix Element: 5,1
Life Stage: Recycling, Disposal
Environmental Concern: Material Choice

If any of the following conditions apply, the matrix element rating is 0:

- Product contains significant quantities of mercury (i.e., mercury relays), asbestos (i.e., asbestos based insulations), or cadmium (i.e., cadmium or zinc plated parts) that are not clearly identified and easily removable.

If all of the following conditions apply, the matrix element rating is 4:

- Material diversity is minimized, the product is easy to disassemble, and all parts are recyclable.

If neither of the above ratings is assigned, complete the checklist below. Assign a rating of 1, 2, or 3 depending on the degree to which the product meets DFE preferences for this matrix element.

- Have materials been chosen and used in light of the desired recycling/disposal option for the product (e.g., for incineration, for recycling, for refurbishment)?
- Does the product minimize the number of different materials that are used in its manufacture?
- Are the different materials easy to identify and separate?
- Is this a battery free product?
- Is this product free of components containing PCBs or PCTs (e.g., in capacitors and transformers)?
- Are major plastic parts free of polybrominated flame retardants or heavy metal-based additives (colorants, conductors, stabilizers, etc)?

Product Matrix Element: 5,2
Life Stage: Recycling, Disposal
Environmental Concern: Energy Use

If any of the following conditions apply, the matrix element rating is 0:

* Recycling/disposal of this product is relatively energy-intensive (compared to other products that perform the same function) due to its weight, construction, and/or complexity.

If all of the following conditions apply, the matrix element rating is 4:

* Energy use for recycling or disposal of this product is minimal.

If neither of the above ratings is assigned, complete the checklist below. Assign a rating of 1, 2, or 3 depending on the degree to which the product meets DFE preferences for this matrix element.

* Is the product designed with the aim of minimizing the use of energy-intensive process steps in disassembly?
* Is the product designed for high-level reuse of materials? (Direct reuse in a similar product is preferable to a degraded reuse.)
* Will transport of products for recycling be energy-intensive because of product weight or volume or the location of recycling facilities?

Product Matrix Element: 5,3
Life Stage: Recycling, Disposal
Environmental Concern: Solid Residues

If any of the following conditions apply, the matrix element rating is 0:

* Product consists primarily of unrecyclable solid materials (such as rubber, fiberglass, and compound polymers).

If all of the following conditions apply, the matrix element rating is 4:

* Product can be easily refurbished and reused and is easily dismantled and 100% reused/recycled at the end of its life. For example, no part of this product will end up in a landfill.

If neither of the above ratings is assigned, complete the checklist below. Assign a rating of 1, 2, or 3 depending on the degree to which the product meets DFE preferences for this matrix element.

- Has the product been assembled with fasteners such as clips or hook-and-loop attachments rather than chemical bonds (gels, potting compounds) or welds?
- Have efforts been made to avoid joining dissimilar materials together in ways difficult to reverse?
- Are all plastic components identified by ISO markings as to their content?
- If product consists of plastic parts is there one dominant (>80% by weight) species?
- Is this product to be leased rather than sold?

Product Matrix Element: 5,4
Life Stage: Recycling, Disposal
Environmental Concern: Liquid Residues

If any of the following conditions apply, the matrix element rating is 0:

- Product contains primarily unrecyclable liquid materials.

If all of the following conditions apply, the matrix element rating is 4:

- Product uses no operating liquids (such as oils, coolants, or hydraulic fluids) and no cleaning agents or solvents are necessary for its reconditioning.

If neither of the above ratings is assigned, complete the checklist below. Assign a rating of 1, 2, or 3 depending on the degree to which the product meets DFE preferences for this matrix element.

- Can liquids contained in the product be recovered at disassembly rather than lost?
- Does disassembly, recovery, and reuse generate liquid residues?
- Does materials recovery and reuse generate liquid residues?

Product Matrix Element: 5,5
Life Stage: Recycling, Disposal
Environmental Concern: Gaseous Residues

If any of the following conditions apply, the matrix element rating is 0:

- Product contains or produces primarily unrecyclable gaseous materials that are dissipated to the atmosphere at the end of its life.

If all of the following conditions apply, the matrix element rating is 4:

- Product contains no substances lost to evaporation/sublimation (other than water) and no volatile substances are used for refurbishment.

If neither of the above ratings is assigned, complete the checklist below. Assign a rating of 1, 2, or 3 depending on the degree to which the product meets DFE preferences for this matrix element.

- Can gases contained in the product be recovered at disassembly rather than lost?
- Does materials recovery and reuse generate gaseous residues?
- Can plastic parts be incinerated without requiring sophisticated air pollution control devices? Specifically avoid the use of plastic parts that:
  - contain polybrominated flame retardants or metal based additives,
  - are finished with polyurethane based paints,
  - are plated/painted with metals.

Conversion Factors and Units of Measurement

The International System of Units (SI) is used throughout this book, in recognition of the common practice in science and technology throughout the world. This choice is somewhat more complex in the case of the automobile and its infrastructure than with some other products, because of the frequent use of such English units as "mile" and "miles per hour." Today's engineer needs to work in SI units, however, so we provide here the conversion factors (Table B1) needed for survival by those for whom the English units are second nature.

Mass is normally given in *kilograms* (i.e., 1000 g). Many of the quantities of mass are large, so prefixes (Table B2) are common. Hence, we have such figures as 2 Pg = 2×10^{15} g. Where the word tonne is used, it refers to the metric ton = 1×10^6 g.

The usual distance unit in this book is the *kilometer*, of which there are 1.6 in a mile. For areas, the *hectare* (= $10^4 \ m^2$) is the usual unit.

The basic unit of energy is the *joule* (= $1 \times 10^7 \ erg$). One will often see the use of the British Thermal Unit (BTU), which is $1.55 \times 10^3 J$. For very large energy use, a unit named the *quad* is common; it is shorthand notation for one quadrillion British Thermal Units. Thus, 1 quad = $1 \times 10^{15} \ BTU = 1.55 \times 10^{18} \ J$.

The environmental impacts of automobiles and infrastructure often involve the emissions of gases and particles. The most common way of expressing the abundance of a gas phase atmospheric species is as a fraction of the number of molecules in a sample of air. The units in common use are *parts per million* (ppm), *parts per billion* (thousand million; ppb), and *parts per trillion* (million million; ppt), all expressed as volume fractions and therefore abbreviated ppmv, ppbv, and pptv to make it clear that one is not speaking of fractions in mass. Particles can be mixtures of solid and liquid, so a measure based on mass replaces that based on volume, the usual units for atmospheric particles being micrograms per cubic meter ($\mu g \ m^{-3}$) or nanograms per cubic meter ($ng \ m^{-3}$).

Table B1   Units and Conversion Factors

Quantity	Name of unit	Unit symbol	English equivalent
Length	Centimeter	cm	0.394 inches
Length	Meter	m	39.37 inches
Length	Kilometer	km	0.621 miles
Mass	Kilogram	kg	2.20 pounds
Volume	Liter	l	0.264 U.S. gallons
Volume	Liter	l	61.0 in^3
Volume	Barrel	bbl	42 U.S. gallons
Area	Hectare	ha	10^4 m^2
Area	Hectare	ha	2.47 acres
Energy	Joule	J	6.45×10^{-4} BTU
Energy	Quad	Q	1×10^{15} BTU
Power	Watt	W	1 J/s
Power	Watt	W	0.293 BTU/hr
Power	Watt	W	1.34×10^{-3} horsepower

Acidity in solution is expressed in pH units, pH being defined as the negative of the logarithm of the hydrogen ion concentration in moles per liter. In aqueous solutions, pH = 7 is neutral at 25 $^o C$; lower pH values are characteristic of acidic solutions, higher values are characteristic of basic solutions.

Table B2   Prefixes for Large and Small Numbers

Power of ten	Prefix	Symbol
+24	yotta	Y
+21	zetta	Z
+18	exa	E
+15	peta	P
+12	tera	T
+9	giga	G
+6	mega	M
+3	kilo	k
-3	milli	m
-6	micro	μ
-9	nano	n
-12	pico	p
-15	femto	f
-18	atto	a
-21	zepto	z
-24	yocto	y

Index